DNA Typing Protocols:
Molecular Biology and Forensic Analysis

Other BioTechniques® Books

Affinity and Immunoaffinity Purification Techniques
T.M. Phillips and B.F. Dickens

Antigen Retrieval Techniques: Immunohistochemistry and Molecular Morphology
S.-R. Shi, J. Gu, and C.R. Taylor (Eds.)

Apoptosis Detection and Assay Methods
L. Zhu and J. Chun (Eds.)

Bioinformatics: A Biologist's Guide to Biocomputing and the Internet
S. Brown

DNA Alterations in Cancer: Genetic and Epigenetic Changes
M. Ehrlich (Ed.)

Gene Cloning and Analysis by RT-PCR
P.D. Siebert and J.W. Larrick (Eds.)

Gene Transfer Methods: Introducing DNA into Living Cells and Organisms
P.A. Norton and L.F. Steel (Eds.)

Immunological Reagents and Solutions: A Laboratory Handbook
B. Damaj

Microarray Biochip Technology
M. Schena (Ed.)

Protein Staining and Identification Techniques
R.C. Allen and B. Budowle

Ribozyme Biochemistry and Biotechnology
G. Krupp and R.K. Gaur (Eds.)

SNP and Microsatellite Genotyping: Markers for Genetic Analysis
A. Hajeer, J. Worthington, and S. John (Eds.)

Viral Vectors: Basic Science and Gene Therapy
A. Cid-Arregui and A. García-Carrancá (Eds.)

DNA Typing Protocols: Molecular Biology and Forensic Analysis

Bruce Budowle, PhD
FBI
Laboratory Division
Washington, DC, USA

Jenifer Smith, PhD
FBI
Laboratory Division
Washington, DC, USA

Tamyra Moretti, PhD
Forensic Science Research Unit
FBI Academy
Quantico, VA, USA

Joseph DiZinno, DDS
FBI
Laboratory Division
Washington, DC, USA

A BioTechniques® Books Publication
Eaton Publishing

Bruce Budowle, PhD
FBI
Laboratory Division
Washington, DC 20535, USA

Jenifer Smith, PhD
FBI
Laboratory Division
Washington, DC 20535, USA

Tamyra Moretti, PhD
Forensic Science Research Unit
FBI Academy
Quantico, VA 22135, USA

Joseph DiZinno, DDS
FBI
Laboratory Division
Washington, DC 20535, USA

Library of Congress Cataloging-in-Publication Data

DNA typing protocols : molecular biology and forensic analysis / Bruce Budowle ... [et al.].
    p. cm.
  Includes bibliographical references and index.
  ISBN 1-881299-23-6
   1. Forensic genetics--Laboratory manuals. 2. DNA fingerprinting--Laboratory
manuals. 3. Polymerase chain reaction--Laboratory manuals. 4.
DNA--Analysis--Laboratory manuals. I. Budowle, Bruce, 1953-

RA1057.5 .D534 2000
614'.1--dc21

                                00-049491

ISBN 1-881299-23-6

Printed in the United States of America

9 8 7 6 5 4 3 2 1

Eaton Publishing
BioTechniques Books Division
154 E. Central Street
Natick, MA 01760
www.BioTechniques.com

Francis W. Eaton: *Publisher and President*
Christine McAndrews: *Managing Editor and Director*
Chanc E VanWinkle: *Project Editor*
Sandy Lamont: *Production Manager*
Ken Strom: *Cover Designer*

Contents

Introduction

1

The analysis of DNA has revolutionized the field of biological identification. The requirement for only minute quantities of tissue, the stability of DNA, and the high degree of assay accuracy and precision have contributed to the evolution of DNA typing. The use of DNA analyses is rapidly altering the manner in which genetic typing is carried out in human parentage, rape, and homicide cases, in the evaluation of mass disasters, in animal poaching, breeding, and population studies, in medical analysis, and in plant patent disputes.

To be effective, techniques that are developed for application-oriented laboratories should be designed to be used readily by practitioners. The limited quantity and/or quality of forensic biological specimens require that DNA typing methods be robust. Protocols tend to be spared of unnecessary steps and optimized, as best as is possible, in the areas of performance time and cost reduction, while maximizing reliability and detection sensitivity. Moreover, the procedures used in forensic applications have been through extensive validation. Validation is a process to assess the ability of defined procedures to reliably obtain desired results, to define conditions that are required to obtain the result, to determine the limitations of the analytical procedure, and to identify aspects that must be monitored and controlled. Validated methods are essential to application-oriented laboratories and provide stability to rapidly evolving fields, such as those that entail molecular biology.

This book provides procedures used by the Federal Bureau of Investigation (FBI) Laboratory that have proven useful for typing reference samples, which typically are in good condition, and specimens found at crime scenes that have been exposed to the environment. Crime scene samples may be degraded and contaminated with materials that may inhibit analytical

DNA Typing Protocols: Molecular Biology and Forensic Analysis
By B. Budowle, J. Smith, T. Moretti, J. DiZinno
©2000 Eaton Publishing, Natick, MA

processes, particularly enzymatic reactions. Because of the demand for robust and reliable procedures, the procedures and practices used in forensic analyses could guide scientists in other disciplines in development and/or employment of valid and reliable methodologies.

Particular topics presented in a specific chapter may apply equally well to protocols in other chapters. For example, polymerase chain reaction (PCR) is first described in Chapter 6, entitled PCR-Based Analyses: Allele-Specific Oligonucleotide Assays, but also is used in the protocols in Chapters 7 through 10. The topics/protocols appear in this book in the sequence that they became established in the forensic field, with the earliest technologies appearing first. At the end of the book are three appendices—one listing quality assurance standards for forensic DNA typing laboratories, one displaying allele frequency data for several genetic markers in some representative population groups, and the last appendix displaying most reagents and supplies, preparation of many solutions, and commercial sources as a guide for the user.

Quality assurance (QA) and quality control (QC) are requisite for ensuring high-quality results. Adherence to using tested reagents, calibrated equipment, and known control samples is necessary to obtain reliable results with confidence from standard operating protocols. Often, when problems arise (with good quality samples) and poor quality or no result is obtained, it is due to not following QA or QC practices. A QA program should be implemented to demonstrate that personnel, equipment, and reagents perform as expected. New lots of critical reagents should be tested before use for routine typing. The use of controls is emphasized. Positive controls (i.e., samples of known type or values) should be used to monitor the performance of the assay. If the types for the positive control sample(s) are incorrect or deviate substantially from the mean, an evaluation of the process is warranted, and the data from that assay may be invalid. Because of the increased sensitivity afforded by the use of the PCR, control measures should be implemented for detecting levels of contamination that may affect the interpretation of an assay. Negative controls are useful to monitor contamination during the PCR.

In 1998, the Director of the FBI issued QA standards for DNA typing laboratories carrying out forensic DNA analyses. These are described in Appendix 1. Although we do not address QA/QC in a specific section, the practices are alluded to throughout the book. In Chapter 9, entitled PCR-Based Analyses: Capillary Electrophoresis for STR Loci Typing and Real Time Fluorescent Detection, we describe QC in greater detail. However, the basic issues of QA/QC apply to all practices in the laboratory.

Appendix 3 provides sources for many of the reagents used. If materials or equipment can be obtained from multiple sources without affecting the assay, then a commercial source may or may not be provided. It is assumed that laboratories are equipped with standard equipment and tools, such as fume hoods, forceps, scissors, centrifuges, etc. These will not be specified in Appendix 3, but will be stated in the protocol.

Not all possible procedures, variations of procedures, or all genetic marker systems are described. Instead, representative examples are provided to demonstrate what can be accomplished and to educate the reader. There is some redundancy for some of the steps in some protocols. This is intentional to facilitate usage of the protocol, whereby each major protocol is complete and will not require additional searching or cross-referencing. The authors believe that the procedures for DNA typing of forensic specimens will be of interest to the general scientific community and may be applicable to research and analyses in other scientific arenas.

Finally, we would like to thank our many collaborators over the years whose efforts also have contributed to the protocols in this book. There are many at the FBI who assisted in developing these protocols, and we are indebted to their efforts and contributions to the application of DNA typing. The following scientists, external to the FBI, particularly have been fruitful collaborators over the years, and we are appreciative of their contributions: Robert C. Allen, Ranajit Chakraborty, Arthur J. Eisenberg, Manfred N. Hochmeister, Leonard Klevan, Jose A. Lorente, Miguel Lorente, Rebecca Reynolds, Antti Sajantila, Mark Stoneking, Sean P. Walsh, and Ray White. We also would like to thank those manufacturers who kindly provided figures and informational support: Fitzco Corporation (Minneapolis, MN, USA), Hitachi Genetic Systems/MiraiBio (Alameda, CA, USA), Life Technologies (Gaithersburg, MD, USA), PE Biosystems (Foster City, CA, USA), and Promega Corporation (Madison, WI, USA). We would also like to thank Claudine Riggio for her photography work and Chanc E VanWinkle (Eaton Publishing) for her input as project editor. Lastly, we thank Steve Weaver (Eaton Publishing) for his patience throughout the effort.

This is publication number 00-02 of the Laboratory Division of the Federal Bureau of Investigation. Names of commercial manufacturers are provided for identification only, and inclusion does not imply endorsement by the Federal Bureau of Investigation.

Sample Collection and Storage

2

Preparation and storage of reference samples are important parts of the analytical process and should not be ignored. Typically, reference samples in a human identification laboratory are either blood or saliva/buccal cells. Peripheral leukocytes are the source of genomic DNA from blood. Liquid blood usually is collected by venipuncture in an EDTA vacutainer tube and can be stored at ambient temperature for short periods of time, but 4°C is preferable. For longer storage, the specimen should be aliquoted (0.7 mL) into 1.5-mL screw cap polypropylene tubes and frozen at -20° or -70°C. Liquid blood may be freeze-thawed as required. This process is, in fact, advantageous to ensure complete lysis of the red cell membranes. Drying blood onto filter paper [particularly FTA paper (Fitzco, Minneapolis, MN, USA)] is an effective form of storage. Prior to short or long term storage, aliquots of blood can be placed onto filter paper, such as 903 Blood Specimen Collection Paper (Schleicher & Schuell, Keene, NH, USA) or FTA paper, and stored dry at room temperature or frozen.

Alternatively, epithelial cells from the inside lining of the cheek (i.e., buccal cells) are a good source of DNA. These saliva samples (or buccal epithelial cells) tend to be collected using buccal swabs. The cells typically are collected onto the swabs by gentle scraping of the inside of the cheek. The swab is then dried or placed in alcohol. This collection of DNA samples is considered less invasive than blood collection, and a phlebotomist is not required for sample collection.

As long as sufficient genomic DNA is available, other sources may potentially serve as a reference sample. Cellular material that can be used as a reference sample may be derived from hair, stamps, envelopes, toothbrushes, and razors.

DNA Typing Protocols: Molecular Biology and Forensic Analysis
By B. Budowle, J. Smith, T. Moretti, J. DiZinno
©2000 Eaton Publishing, Natick, MA

All reference samples should be considered biohazardous, and proper handling precautions should be taken. When appropriate, wear gloves, masks, and/or work under a hood.

2.1 REFERENCE SAMPLE PREPARATION OF DRIED BLOODSTAINS FROM COAGULATED BLOOD

Note: Universal precautions should be practiced when handling biological materials.

Materials and Reagents

- Cotton sheeting
- Glass tissue homogenizer
- Pasteur pipet

- 10% bleach

Procedure

1. From cotton sheeting, cut a piece approximately 3×6 cm and attach an information label. Attach the cutting to a support.
2. Place a disposable, leak-resistant mat beneath the cutting.
3. Remove the blood tube stopper and place it on a disposable surface.
4. Place a glass tissue homogenizer in a rack capable of holding it upright in a stable manner.
5. Slowly transfer the contents of the clotted blood tube to the glass homogenizer.
6. Insert the glass pestle and gently homogenize the clotted material and serum. Rapid homogenization strokes will lead to excessive foaming and protein denaturation.
7. Remove the pestle and place it on a clean, disposable surface.
8. With a Pasteur pipet, transfer a portion of the homogenized clot and serum onto the cotton cutting. Be careful to add the liquid to the cotton cutting slowly to avoid drips and runoffs.
9. Transfer unused homogenate back into the original tube and replace the tube's stopper. Tape the stopper to seal.
10. Treat the Pasteur pipet, the glass homogenizer, and pestle with 10% bleach solution. Discard the Pasteur pipet in a biohazard container. Wash the glass homogenizer with detergent and rinse copiously with water.
11. Permit the bloodstain to dry thoroughly at room temperature.
12. Place the dried stain in a labeled envelope and store at -20°C until used.

2.2 PREPARATION OF DRIED BLOODSTAINS FROM ANTICOAGULATED BLOOD

Materials and Reagents

- Cotton sheeting
- Glass tissue homogenizer
- Pasteur pipet

- 10% bleach

Procedure

1. From cotton sheeting, cut a piece approximately 3×6 cm and attach an information label. Attach the cutting to a support.

2. Place a disposable, leak-resistant mat beneath the cutting.

3. Invert the tube several times to mix the contents.

4. Remove the stopper from the tube and place it on a disposable surface.

5. With a Pasteur pipet, transfer a portion of the liquid blood onto the cotton sheeting cutting. Be careful to add the liquid to the cutting slowly to avoid drips and runoffs.

6. Treat the Pasteur pipet, the glass homogenizer, and pestle with 10% bleach solution. Discard the Pasteur pipet in a biohazard container. Wash the glass homogenizer with detergent and rinse copiously with water.

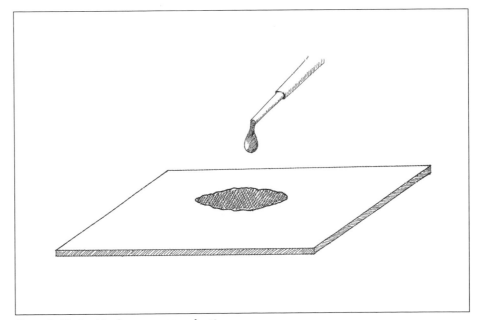

Figure 1. Placing blood spot on cotton sheeting or paper.

7. Permit the bloodstain to dry thoroughly at room temperature.

8. Place the dried stain in a labeled envelope and store at 4°C or -20°C.

2.3 FTA PAPER PROTOCOL

FTA paper provides a means for collecting, transporting, and/or storing blood samples and possibly other biological specimens. Blood can be deposited directly from a fingerprick onto the paper, or blood from an EDTA vacutainer tube can be deposited by pipet onto the FTA paper. Cell membranes are lysed in the FTA matrix, and the DNA from the cells becomes entangled in the paper. The filter paper entraps DNA such that it is essentially immobilized. Thus, proteins and cellular debris can be washed away from the sample, and yet the DNA remains intact and is purified within the paper. The washed paper, containing the DNA, is the conduit for sample analysis in restriction fragment-length polymorphism (RFLP) or PCR-based assays. The FTA paper is impregnated with denaturants that prevent bacterial and fungal growth, as well as oxidative and UV damage. Thus, blood and other samples stored on FTA paper will be stable for long time periods and may be stored at ambient temperature.

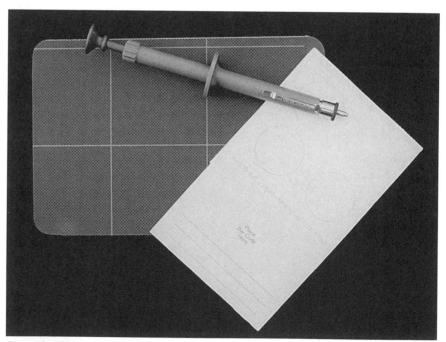

Figure 2. FTA paper is useful for long-term storage of samples, because of anti-bacterial and anti-fungal properties. A sample can be punched from the paper, washed, and used directly in a PCR amplification.

Materials and Reagents

- FTA paper

Procedure

1. Spot liquid blood sample (5–100 µL) directly onto FTA paper and allow to dry at least 1 h.
2. Place paper in a protective envelope and store in the dark at ambient temperature (or store refrigerated or frozen if desired).

2.4 BUCCAL SWAB PROTOCOL

Buccal cells can be a good source of genomic DNA, and the sampling process is less invasive than that for drawing blood.

Materials and Reagents

- C.E.P. SWAB™ (Life Technologies, Gaithersburg, MD, USA)
- Foam tip applicator — Gene Guard Swab (Life Technologies)
- Isopropyl alcohol

Procedure

1. Place C.E.P. SWAB in mouth.
2. Firmly move the swab around inside of left cheek. Use both sides of swab during swabbing.
3. Allow C.E.P. SWAB to air dry or place swab in alcohol.
4. Repeat steps 1 through 3 with new C.E.P. SWAB for inside of right cheek.
5. Store C.E.P. SWAB as appropriate.
6. If foam tip swab (Gene Guard Swab) is used, swab only with one side of swab foam applicator.
7. While still moist, firmly press the swab, do not rub the foam tip, on the FTA paper.
8. Allow FTA paper to dry at least 30 minutes.
9. Secure and store FTA paper as appropriate.

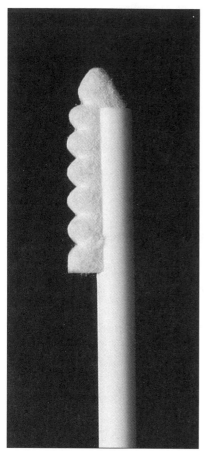

Figure 3. C.E.P. Sᴡᴀʙ for collecting buccal cells. The brush-like head can be injected into alcohol for sample storage, or it can be air-dried and stored.

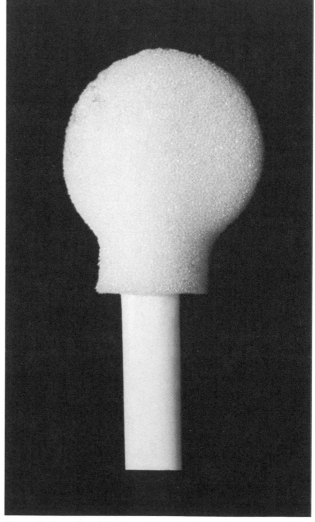

Figure 4. Gene Guard Swab used for collecting buccal cells.

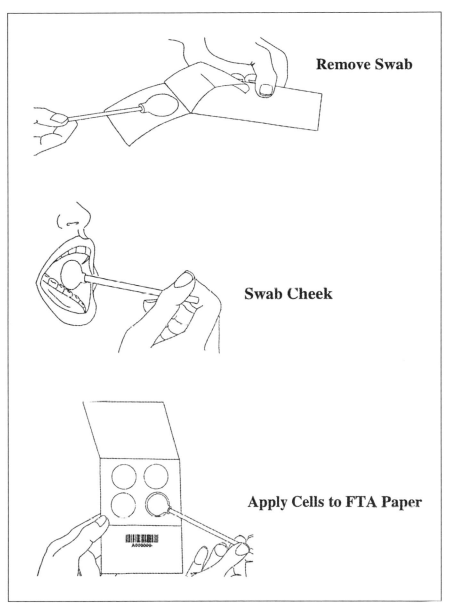

Remove Swab

Swab Cheek

Apply Cells to FTA Paper

Figure 5. Collection of buccal cells using a foam applicator swab. After swabbing, the swab is pressed against FTA paper to place the buccal cells on the storage medium (kindly provided by Fitzco Corporation).

SUGGESTED READING

1. **Belgrader, P., S.A. Del Rio, K. Turner, M.A. Marino, K.R. Weaver, and P.E. Williams.** 1995. Automated DNA purification and amplification from blood-stained cards using a robotic workstation. BioTechniques *19*:426-432.

2. **Belgrader, P. and M.A. Marino.** 1997. Automated sample processing using robotics for genetic typing of short tandem repeat polymorphisms by capillary electrophoresis. Lab. Robot. Automat. *9*:3-7.

3. **Bever, R.A. and M.A. DeGuglielmo.** 1994. Development and validation of buccal swab collection method for DNA testing. Adv. Forensic Haemogent. *5*:199-201.

4. **Burgoyne, L. and C. Rogers.** 1997. Bacterial typing: storage and processing of stabilized reference bacteria for polymerase chain reaction without preparing DNA—an example of an automatable procedure. Anal. Biochem. *247*:223-227.

5. **Fisher, B.J., A. Svensson, and O. Wendal.** 1987. Blood and other biological evidence, p. 197-198. *In* Techniques of Crime Scene Investigation. Elsevier, New York.

6. **Maniatis, T., E.F. Fritsch, and J. Sambrook.** 1982. Molecular Cloning: A Laboratory Manual. CSH Laboratory Press, Cold Spring Harbor, NY.

7. **Sambrook, J., E.F. Fritsch, and T. Maniatis.** 1989. Molecular Cloning: A Laboratory Manual, Vols. 1, 2, and 3. 2nd ed. CSH Laboratory Press, Cold Spring Harbor, NY.

Extraction of DNA

3

DNA can be extracted from most types of biological material found at a crime scene. The success of DNA typing relies on the isolation of DNA of sufficient quantity, quality, and purity. Depending on the typing procedure, the required quantity (i.e., the amount of retrievable DNA) and quality (i.e., the length of the DNA molecules) of DNA can vary widely. Purity of the DNA extract refers to a quality of cleanliness, such that subsequent analytical assays can be carried out effectively. Generally, DNA extraction protocols that overcome, remove, or dilute enzymatic inhibitors are desirable. Although many assays are similarly based, there is a tremendous number of variations, and many can work effectively.

Ideally, extraction protocols should be simple and inexpensive to perform. Procedures should be suitable for extraction of DNA from small liquid blood samples, bloodstains, and other tissues. Procedures suitable for extraction of DNA from small liquid blood samples and bloodstains are based on standard DNA extraction methods. Standard DNA extraction procedures may entail (*i*) organic solvent, (*ii*) salting out methods, or (*iii*) cation exchange resins such as Chelex 100 (Bio-Rad Laboratories, Hercules, CA, USA). Organic and salting out methods are compatible with various typing procedures, including restriction fragment-length polymorphism (RFLP) typing and polymerase chain reaction (PCR)-based analyses and are routinely used.

The organic extraction procedure yields highly purified DNA. The process involves (*i*) lysis of the red cells with SSC buffer (NaCl, sodium citrate), (*ii*) centrifugation, (*iii*) lysis of the pelleted white cells, and (*iv*) digestion of proteinaceous materials by incubation in a sodium acetate, sodium dodecyl sulfate (SDS), and proteinase K solution. DNA released into the solution is extracted with phenol to remove proteinaceous material, and chloroform is present to remove residual phenol. The DNA is precipitated from the aqueous

DNA Typing Protocols: Molecular Biology and Forensic Analysis
By B. Budowle, J. Smith, T. Moretti, J. DiZinno
©2000 Eaton Publishing, Natick, MA

layer by the addition of cold ethanol and salt and by subjecting the solution to centrifugation. Considerable care must be exercised not to disturb the DNA pellet when removing the ethanol. Residual ethanol may be removed under vacuum in a specially designed centrifuge. The resultant pellet should be slightly moist to ensure resuspension in Tris-EDTA (TE) buffer.

Instead of precipitating the DNA with alcohol, salts can be removed by dialysis. The simplest way to achieve dialysis is by centrifugation through a filter device, such as an Amicon® Microcon™ 100 filter unit (Millipore, Bedford, MA, USA). If the DNA is to be amplified by PCR, as opposed to restriction analysis, a procedure involving Microcon 100 dialysis of the aqueous layer can be followed.

Nonorganic extraction methods that use NaCl or LiCl to salt out protein have been described. Because salts may not be adequately removed using those procedures, problems may occur with restriction endonuclease digestion and band shifting (alteration of DNA mobility). However, with proper care these methods can produce satisfactory results.

The above procedures are compatible with both RFLP and PCR-based procedures, because the isolated DNA is double stranded. The Chelex method is only compatible with PCR-based assays. Basically, Chelex 100 is an ion-exchange resin that binds cations—removing them from solution—that may inhibit the PCR process. The Chelex protocol entails placing a small sample in 5% (wt/vol) Chelex and incubating the sample at 56°C for 30 minutes. After incubation, the sample is boiled for 8 minutes and centrifuged. A portion of the supernatant is used for the PCR. Boiling denatures the DNA into the single strand format, rendering the DNA unsuitable for restriction enzyme digestion (i.e., RFLP typing). However, chelex-extracted DNA is compatible with the PCR. The DNA extract used for the PCR should not contain residual Chelex; Chelex will bind magnesium ions, which are needed for polymerase activity.

A method of purification that is particularly useful for reference samples is the washing away of cellular material from DNA immobilized in FTA paper (Fitzco, Minneapolis, MN, USA). Blood is spotted onto FTA paper, the cells are lysed, the DNA released from the cells is immobilized in the FTA matrix, and the sample is dried. Heme and other cellular debris, which may inhibit enzymatic activity in subsequent assays, are simply washed away. The washed, immobilized DNA on a circular punch (approximately 1.2 mm in diameter) can be used directly as a DNA template source.

When an assay demands the highest quality DNA, the method of choice for extraction of DNA is the organic method. Although the protocol generally is more time-consuming and laborious, the purity of the extracted DNA usually is higher than other methods. More protein is removed from the DNA molecule during organic extraction compared with other methods, and, when coupled with a filtration wash, many enzymatic inhibitors are removed.

The majority of biological evidence analyzed in a United States forensic laboratory is derived from rape cases. The evidence from rape cases, such as vaginal swabs and stained clothing, most often contains nucleated cells from the male contributor (i.e., predominately sperm) and the female victim (i.e., epithelial cells). Elucidation of the individual contributors' DNA profiles can, at times, be complicated in these mixtures. However, sperm cells can be separated from other cells during extraction. Sperm cell membranes contain thiol-rich proteins and are resistant to cell lysis in the absence of a chemical known as a reducing agent. The lack of use of a reducing agent results in a differential lysis in which nonsperm cells preferentially burst. The nonsperm DNA is released into the supernatant, creating a fraction that predominately contains DNA from the epithelial cells from the female contributor. Subsequently, in the presence of a reducing agent, DNA from the intact sperm can be extracted separately from the female cells. The process purifies and enriches each fraction, so that DNA profiles from both male and female contributors in a rape case sample can be more readily identified.

Several approaches for extracting DNA are described so that the reader may choose a method(s) that suits his/her needs. The methods—organic with alcohol precipitation, organic with dialysis washing, chelex extraction, and salting out—provide a means of encompassing a wide variety of tissue samples.

The following general equipment and supplies are used:

Pipettors: adjustable 0.5–10 µL, 2–20 µL, 20–200 µL, and 200–1000 µL
Microcentrifuge tube racks
Microcentrifuge tubes
Microcentrifuges
Centrifuge for Amicon concentration (e.g., benchtop clinical centrifuge)
Scalpel and disposable scalpel blades
Aerosol-resistant pipet tips
Electric or battery-charged pipet aid
Sterile, disposable serological pipets
Sterile, disposable 15 and 50 mL plastic screw cap tubes

Refrigerator, freezers (-20° and -70° to -80°C)
Heat/stir plate
Sterile laboratory glassware
pH meter
Disposable gloves
Protective clothing and safety goggles
Forceps
Stationary water baths (37° and 56°C)
Shaking water bath (50°C)
Orbital shaker
Rocking platform
Vacuum source

During the isolation of DNA from evidentiary materials, only the items from a single case are present, and only one item from the case is opened and processed. This practice is designed to prevent the misidentification of any of the items involved in the submitted case and to minimize contamination.

3.1 ORGANIC EXTRACTION/ISOLATION OF DNA FROM LIQUID BLOOD SAMPLES—ALCOHOL PRECIPITATION METHOD

Materials and Reagents

- EDTA Vacutainer™ tubes
- Sarstedt tubes (Newton, NC, USA)
- Eppendorf® tube
- Speed-Vac® Centrifuge (Savant Instruments, Holbrook, NY, USA)

- 1× SSC
- 0.2 M sodium acetate
- 10% SDS
- Proteinase K (20 mg/mL in H_2O)

- Phenol/chloroform/isoamyl alcohol (PCIA) (25:24:1)
- Absolute alcohol
- TE buffer
- 2.0 M sodium acetate
- 70% EtOH

Procedure

1. Liquid blood specimens should be collected in EDTA vacutainer tubes. Gently mix the contents before removing aliquots. Liquid blood can be stored at 4°C for at least five days. Otherwise, it is desirable to aliquot and freeze the blood. Freeze the blood at -80°C in 700-µL aliquots in Sarstedt tubes (screw caps).

2. Add 800 µL 1× SSC to 700 µL of liquid blood or thawed blood (if previously frozen), mix, and centrifuge at maximum speed for 1 min in a microcentrifuge.

3. Remove 1.0 mL of supernatant and discard the supernatant into a disinfectant solution.

4. Add 1.0 mL 1× SSC to the tube, vortex-mix, and centrifuge for 1 min. Remove and discard as much of the supernatant as possible. Do not disturb the pellet.

5. To the pellet, add 375 µL 0.2 M sodium acetate, 25 µL 10% SDS, and 5 µL proteinase K (20 mg/mL). Vortex-mix the contents briefly (approximately 1 s) and incubate at 56°C for 1 h.

   Note: Processing of the allelic control specimens (see Chapter 4, entitled Restriction Fragment-Length Polymorphism Typing of Variable Number Tandem Repeat Loci) begins at this step.

6. Add 120 µL PCIA to the tube and vortex-mix for 30 s. Perform this step in the fume hood.

   CAUTION: PCIA is an irritant and is toxic. Its use should be confined to a fume hood. Gloves and a mask should be worn.

   Note: If necessary, additional organic extractions may be performed prior to the purification steps.

7. Centrifuge for 2 min.

8. The aqueous layer (i.e., top layer), which contains DNA, is carefully removed and placed in a new 1.5-mL Eppendorf tube. Try not to remove the layer of denatured protein that collects at the interface of the aqueous and organic layers. Discard the original tube and PCIA in an appropriate waste container housed in the fume hood.

9. Add 1.0 mL cold, absolute EtOH to the aqueous layer. Mix by inverting the tube a couple of times. Place the tube at -20°C for 15 min.

Note: Samples can be stored at 4°C at this stage. Centrifuge for 2 min.

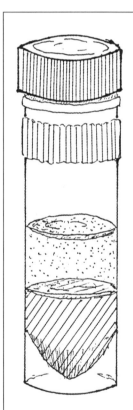

Figure 1. A tube containing the aqueous and organic phases of the extraction. The aqueous phase (i.e., the DNA-containing phase) is on top. Do not disturb the layer between the phases when removing the aqueous phase.

10. Decant and discard the supernatant. Remove any additional alcohol using a micropipettor. Be careful to avoid drawing any of the pellet up into the pipet tip.

11. Add 180 µL TE buffer to the pellet and vortex-mix briefly.

12. Incubate the contents at 56°C for 10 min.

13. Add 20 µL 2.0 M sodium acetate and mix by hand for 5 s.

14. Add 500 µL cold, absolute EtOH and gently mix by hand until the solution is homogeneous.

15. Centrifuge for 1 min.

16. Decant and discard the supernatant.

17. Wash the pellet with 1.0 mL ambient temperature 70% EtOH.

18. Centrifuge for 1 min and decant and discard the supernatant. Remove as much EtOH as possible using a micropipet.

19. Place tubes in a Speed-Vac centrifuge to remove excess EtOH. This usually takes about 10 min.

20. Resuspend the DNA by adding 200 µL TE buffer. Mix and incubate the contents at 56°C overnight.

21. The next day, vortex-mix the tubes for 10 to 30 s.

22. The DNA is ready for quantification.

3.2 ORGANIC EXTRACTION/ISOLATION OF DNA FROM BODY FLUID STAINS—ALCOHOL PRECIPITATION METHOD

Materials and Reagents

- EDTA Vacutainer tubes
- Sarstedt tubes
- Eppendorf tubes
- Speed-Vac Centrifuge

- 1× SSC
- 0.2 M sodium acetate
- 10% SDS
- Proteinase K (20 mg/mL)
- PCIA (25:24:1)
- Absolute alcohol
- TE buffer
- 2.0 M sodium acetate
- 70% EtOH

Procedure

1. Cut the stained material into small pieces and place the pieces into a 1.5-mL tube that has a depression in the cap (Sarstedt).

2. Add 400 μL stain extraction buffer (SEB) and 10 μL proteinase K. Mix and centrifuge for 2 s to force the cut materials into the extraction solution.

3. Incubate at 56°C overnight.

4. Punch a hole in the lid and place the cutting pieces in the lid, centrifuge for 5 min, and remove the cutting pieces and the cap. Place a new cap on the tube.

5. Add 500 μL PCIA. This step is done in the fume hood. Vortex-mix the tube for 20 s to achieve a milky emulsion. Centrifuge the tube at high speed for 2 min in a table-top centrifuge.

   CAUTION: PCIA is an irritant and is toxic. Its use should be confined to a fume hood. Gloves and a mask should be worn.

   Note: If desired, additional organic extractions may be performed prior to the purification steps.

6. The aqueous layer (i.e., top layer), which contains the DNA, is carefully removed and placed in a new 1.5-mL microcentrifuge tube. Try not to remove the layer of denatured protein that collects at the interface of the aqueous and organic layers. Discard the original tube and PCIA in an appropriate waste container housed in the fume hood.

7. Add 1.0 mL cold, absolute EtOH to the aqueous layer.

8. Mix by hand and place the tube at -20°C for 30 min.

9. Centrifuge for 15 min.

10. Decant and discard the alcohol.

11. Wash the pellet with 1.0 mL room temperature 70% EtOH.

12. Centrifuge for 5 min.

13. Decant and discard the alcohol. Remove as much as possible of the remaining alcohol with a micropipet.

14. Place the tube in Speed-Vac for 15 to 30 min to remove residual EtOH.

15. Resolubilize the DNA in 36 μL TE buffer at 56°C for at least 2 h. DNA can be allowed to resolubilize overnight at 56°C.

3.3 ORGANIC EXTRACTION/ISOLATION AND SEPARATION OF SPERM CELL DNA AND VAGINAL CELL DNA—ALCOHOL PRECIPITATION METHOD

Materials and Reagents

- EDTA Vacutainer tubes
- Sarstedt tubes
- Eppendorf tubes
- Speed-Vac Centrifuge

- 1× SSC
- Tris/EDTA/NaCl (TNE)
- 20% sarkosyl
- 0.39 M dithiothreitol (DTT)
- Proteinase K (20 mg/mL in H_2O)
- PCIA (25:24:1)
- Absolute ethanol
- TE buffer
- 2.0 M sodium acetate
- 10% SDS
- 70% EtOH

Procedure

1. Place the semen-containing material (i.e., cut portion of a swab, cuttings of stained material) into a 1.5-mL tube.

2. Add 400 µL TNE, 25 µL 20% sarkosyl, 75 µL water, and 5 µL proteinase K (20 mg/mL). Mix tube contents and incubate at 37°C for 2 h.

3. Punch a hole in the cap of the tube. Place the material into the cap and centrifuge the tube for 5 min.

4. Remove the supernatant and place it into a new 1.5-mL tube. This is the fraction that contains DNA from lysed cells (also known as the female DNA fraction). Remove the substrate material. Be careful not to disturb the pellet in the bottom of the original tube.

5. Place a new cap on the original tube.

6. Add to the pellet in the original tube 150 µL TNE, 50 µL 20% sarkosyl, 40 µL 0.39 M DTT, 150 µL water, and 10 µL proteinase K, gently mix the tube contents, and incubate at 37°C for 2 h.

7. Add 500 µL PCIA to all tubes. This step must be done in the fume hood. Vortex-mix the tube for 20 s to achieve a milky emulsion in the tube. Spin the tube at high speed in a table-top centrifuge for 2 min.

   CAUTION: PCIA is an irritant and is toxic. Its use should be confined to a fume hood. Gloves and a mask should be worn.

   Note: If desired, additional organic extractions may be performed prior to the purification steps.

8. The aqueous layer (i.e., top layer), which contains the DNA, is carefully removed and placed in a new 1.5-mL microcentrifuge tube. Try not to remove the layer of denatured protein that collects at the interface of the aqueous and organic layers. Discard the original tube and PCIA in an appropriate waste container housed in a fume hood.

9. Add 1.0 mL cold absolute EtOH to the aqueous layer.

10. Mix by hand and place the tube at -20°C for 30 min.

11. Centrifuge 15 min.

12. Decant and discard the alcohol.

13. Wash pellet with 1.0 mL room temperature 70% EtOH.

14. Centrifuge for 5 min.

15. Decant and discard the alcohol. Remove remaining alcohol with a micropipet.

16. Place the tube in Speed-Vac for 15 to 30 min to remove residual EtOH.

17. Resolubilize the DNA in 36 μL TE at 56°C for at least 2 h. DNA can be allowed to resolubilize overnight at 56°C.

3.4 ORGANIC EXTRACTION OF DNA FROM BLOODSTAINS AND SALIVA STAINS—MICROCON 100 WASHING

Liquid blood samples can be made into bloodstains. Stains should be air-dried before being stored in individual envelopes at -20°C until used.

Materials and Reagents

- Microcentrifuge tube (Catalog No. 3214; Corning Costar, Acton, MA, USA)
- Wooden stick applicator
- Spin-X basket insert (Catalog No. 9301; Corning Costar)

- SEB
- Proteinase K
- PCIA (25:24:1)
- TE buffer

Procedure

1. Place the bloodstain (approximately 3 × 3 mm) or saliva stain in a 2.2-mL microcentrifuge tube.

2. To the sample, add 300 μL SEB and 2 μL proteinase K solution. Vortex-mix on low speed for 1 s and centrifuge in a microcentrifuge for 2 s to force the cutting into the extraction solution.

3. Incubate the tube at 56°C overnight.

4. Centrifuge in a microcentrifuge for 2 s to force liquid into the bottom of the tube.

5. Using a wooden applicator stick, transfer the cutting into a Spin-X basket insert. Place the basket insert into the tube containing the stain extract. Spin in a microcentrifuge for 5 min. (This step is performed if a cutting is used for extraction).

6. Remove and discard the basket insert plus cutting into a container suitable for biohazardous materials.

7. In a fume hood, add 300 μL PCIA to the stain extract. Vortex-mix (low speed) the mixture briefly to attain a milky emulsion. Spin the tube in a microcentrifuge for 3 min.

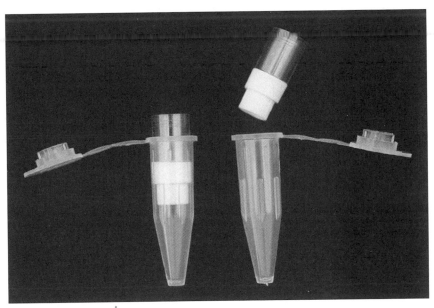

Figure 2. Microcon device.

CAUTION: PCIA is an irritant and is toxic. Its use should be confined to a fume hood. Gloves and a mask should be worn.

Note: If desired, additional organic extractions may be performed prior to the purification steps.

8. Assemble and label a Microcon 100 unit for each sample (see Microcon Operating Instructions). Prepare the Microcon 100 concentrators by adding 100 μL of distilled deionized water to the filter side (top) of each concentrator.

9. To a Microcon 100 concentrator, add 100 μL TE buffer. Transfer the aqueous phase from the tube in step 7 to the concentrator. Avoid pipetting organic solvent from the tube into the concentrator.

10. Place a cap on the concentrator and centrifuge in a microcentrifuge at 500× *g* for 10 min.

11. Carefully remove the concentrator unit from the assembly and discard the fluid from the filtrate cup. Return the concentrator to the top of the filtrate cup.

12. Add 200 μL TE buffer to the concentrator. Replace the cap and centrifuge the assembly in a microcentrifuge at 500× *g* for 10 min.

13. Remove the cap and add a measured volume of TE buffer that is between 20 and 200 μL to the concentrator. Remove the concentrator from the filtrate cup and carefully invert the concentrator into a labeled retentate cup. Discard the filtrate cup.

Note: Final recovery volumes following purification generally range from 20 μL for evidentiary samples to 200 μL for known reference samples.

14. Centrifuge the assembly in a microcentrifuge at 500× *g* for 5 min.

15. Discard the concentrator. Cap the retentate cup.

16. Store samples at 4°C or frozen. Prior to the use of samples after storage, they should be vortex-mixed briefly and centrifuged in a microcentrifuge for 5 s.

3.5 ORGANIC EXTRACTION OF DNA FROM VAGINAL SWABS OR SEMEN STAINS—MICROCON 100 WASHING

Vaginal swabs or semen stains should be air-dried and stored at -20°C until used.

Materials and Reagents

- Microcentrifuge tube
- Wooden stick applicator
- Spin-X basket insert

- TNE
- 20% sarkosyl
- Proteinase K
- Sperm wash buffer
- 1.0 M DTT
- PCIA (25:24:1)
- TE

Procedure

1. Using a clean surface for each swab, dissect the swab material from the applicator stick and place it into a 2.2-mL microcentrifuge tube.

2. To the sample, add 400 µL TNE, 25 µL 20% sarkosyl, 75 µL water, and 1 µL proteinase K. Vortex-mix for 1 s and centrifuge in a microcentrifuge for 2 s to force the material into the extraction solution.

3. Incubate at 37°C for 2 h.

4. Using a wooden applicator stick, transfer the swab material into a Spin-X basket insert. Place the basket insert into the tube containing the stain extract. Centrifuge in a microcentrifuge at maximum speed for 5 min.

5. Remove the basket insert from the extract tube. Remove the swab material from the basket and place in a clean microcentrifuge tube (freeze if storage is required or discard the swab material).

6. While being very careful to not disturb any pelleted material, remove the supernatant fluid from the extract and place it into a new labeled tube. THIS SUPERNATANT IS THE LYSED CELL FRACTION (ALSO KNOWN AS THE FEMALE FRACTION). ANALYSIS OF THE FEMALE FRACTION RESUMES AT STEP 11. THE PELLET REMAINING IN THE TUBE IS THE SPERM CELL PELLET.

7. Wash the cell pellet by resuspending it in 500 µL sperm wash buffer, vortex-mixing the suspension briefly, and centrifuging the tube in a microcentrifuge at maximum speed for 5 min. Remove and discard the supernatant, being careful not to disturb the cell pellet.

8. Repeat step 7 two additional times for a total of three washes of the cell pellet.

9. To the tube containing the washed pellet, add 150 µL TNE, 50 µL 20% sarkosyl, 7 µL 1.0 M DTT, 150 µL water, and 2 µL proteinase K. Close the tube caps and vortex-mix for 1 s and centrifuge in a microcentrifuge for 2 s to force all fluid and material to the bottoms of the tubes.

10. Incubate at 37°C for 2 h.

11. To the tube containing the cell pellet and to the tube containing the female fraction, add 400 µL PCIA. Vortex-mix (low speed) the mixture briefly to attain a milky emulsion. Centrifuge the tube in a microcentrifuge for 3 min.

   CAUTION: PCIA is an irritant and is toxic. Its use should be confined to a fume hood. Gloves and a mask should be worn.

   Note: If desired, additional organic extractions may be performed prior to the purification steps.

12. Assemble a Microcon 100 concentrator unit. To the top of the concentrator, add 100 µL TE buffer. Transfer the aqueous phase from the tube in step No. 11 to the top of the concentrator. Avoid pipetting organic solvent from the tube into the concentrator.

13. Place a cap on the concentrator and centrifuge in a microcentrifuge at 500× *g* for 10 min.

14. Carefully remove the concentrator unit from the assembly and discard the filtrate fluid from the filtrate cup. Return the concentrator to the top of the filtrate cup.

15. Add 200 µL TE buffer to the concentrator. Replace the cap and centrifuge the assembly in a microcentrifuge at 500× *g* for 10 min.

16. Remove the cap and add a measured volume of TE buffer between 20 and 200 µL to the concentrator. Remove the concentrator from the filtrate cup and carefully invert the concentrator onto a labeled retentate cup. Discard the filtrate cup.

   Note: Final recovery volumes following purification generally range from 20 µL for evidentiary samples to 200 µL for known reference samples.

17. Centrifuge the assembly in a microcentrifuge at 500× *g* for 5 min.

18. Discard the concentrator. Cap the retentate cup.

19. Estimate the quantity of DNA in the sample by slot blot hybridization.

20. After quantification, the sample can be amplified.

21. Store samples at 4°C or frozen. Prior to the use of samples after storage they should be vortex-mixed briefly and spun in a microcentrifuge for 5 s.

3.6   ENVELOPE FLAPS OR STAMPS

Procedure

1. Carefully open the envelope flap or remove the stamp using steam and clean forceps. Using a sterile cotton swab moistened in sterile deionized water, swab the gummed envelope flap or stamp. Cut the cotton swab from the stick and transfer the cotton to a 2.2-mL microcentrifuge tube.

OR

1. With a new, sterile scalpel blade, carefully cut one-half of the stamp from the envelope.

    Note: This means that the portion of the envelope to which the stamp is adherent will be removed with the stamp. For envelope flaps, use sterile scissors or a scalpel blade to cut out a portion of the flap approximately 1 cm^2 in size. Cut the stamp or the flap into smaller pieces and place the pieces into a 2.2-mL microcentrifuge tube.

2. Go to step No. 2 of organic extraction procedure (entitled Organic Extraction/Isolation of DNA from Body Fluid Stains—Alcohol Precipitation Method).

3.7   CIGARETTE BUTTS

Procedure

1. Collect a portion of the filter and/or paper of the cigarette butt in the area that would have been in contact with the mouth. Cut it into smaller pieces and place into a 2.2-mL microcentrifuge tube.

2. Go to step No. 2 of organic extraction procedure (entitled Organic Extraction/Isolation of DNA from Body Fluid Stains—Alcohol Precipitation Method).

3.8   SOFT TISSUES

Tissues should be stored at -20°C until used.

Procedure

1. Place a portion of the tissue on a clean surface. Remove a piece of tissue approximately 1 cm^2.

2a. Mince tissue into small pieces and place in a 2.2 or a 1.5-mL microcentrifuge tube.

OR

2b. Transfer tissue to the SPEX 6700 Freezer/Mill (see procedure entitled

Organic Extraction/Isolation of DNA from Body Fluid Stains—Alcohol Precipitation Method) and process following the equipment instructions. Place the processed specimen in a 2.2 or a 1.5-mL microcentrifuge tube.

3. Go to step No. 2 of organic extraction procedure (procedure entitled Organic Extraction/Isolation of DNA from Body Fluid Stains—Alcohol Precipitation Method).

3.9 HAIRS

Protocol for Unmounted Hairs With Sheath Material—for Nuclear DNA Typing

Materials and Reagents

- 100% ethanol
- Sterile deionized water

Procedure

1. While holding the hair specimen with forceps, rinse it thoroughly in 100% ethanol. Follow the ethanol rinse with a thorough rinse in sterile, deionized water.

2. Place at least 1 cm of the hair root end into a 2.2-mL microcentrifuge tube. Alternatively, the entire hair can be placed into the tube.

3. Go to step No. 2 of organic extraction procedure (entitled Organic Extraction/Isolation of DNA from Body Fluid Stains—Alcohol Precipitation Method).

Protocol for Slide-Mounted Hair Specimens

Materials and Reagents

- Xylene

Procedure

1. Loosen the slide coverslip by carefully pipetting xylene around the coverslip edges. If the coverslip does not loosen, the entire slide can be placed into a petri dish and covered with xylene for 1 or more hours until the coverslip has loosened.

2. After removal of the coverslip, remove the hair and rinse it thoroughly with xylene.

3. Continue processing at step No. 1 of the procedure for unmounted hair specimens.

3.10   BONE

Prior to characterization of DNA from bone samples, the bone should be examined by a forensic pathologist and/or a forensic anthropologist. All bones should be photographed. If soft tissue is present, DNA may be extracted from the soft tissue.

Note: For mtDNA analysis, distilled deionized water is listed in the following protocols. If contaminating DNA (above acceptable levels) is due to the distilled deionized water, then the use of high purity water (VWR; Catalog No. 72060-094) is recommended.

Protocol When Bone Marrow is Accessible

Materials and Reagents

- Stryker saw
- Dremel
- Microcentrifuge tube
- Microcon 100 unit

- SEB
- Proteinase K
- 10% bleach
- PCIA (25:24:1)
- Distilled, deionized water

Procedure

1. Prepare a reagent blank by adding 300 µL of SEB and 2 µL of 600 U/mL of proteinase K to a sterile 1.5-mL microcentrifuge tube.

2. Cut the bone using a stryker saw or a dremel with a separating disk to a piece measuring approximately 2×5 cm. The separating disk should be used for only one sample and discarded.

3. A dremel with a small round bur may be used if the bone has accessible marrow space. The bur should be cleaned between each sample with alcohol swabs and a 10% bleach solution and rinsed thoroughly with sterile water.

4. With the dremel, remove bone from within the marrow space and collect the bone powder on aluminum foil or wax paper.

5. Carefully pour the bone powder into a sterile 1.5-mL microcentrifuge tube.

6. Add 300 µL of SEB and 2 µL of 600 U/mL of proteinase K.

7. Briefly vortex-mix and centrifuge the tubes in a microcentrifuge to force the bone fragments into the SEB.

8. Incubate at 56°C for a minimum of 2 h and a maximum of 24 h.

9. Remove the tubes from the water bath and briefly centrifuge in a micro-centrifuge.

10. Add 300 µL of PCIA (25:24:1) to each tube.

    CAUTION: PCIA is an irritant and is toxic. Its use should be confined to a fume hood. Gloves and a mask should be worn.

11. Vortex-mix for 30 s to attain a milky emulsion, then centrifuge the tubes in a microcentrifuge 3 min at $10\,000\times g$.

12. Assemble and label a Microcon 100 unit for each sample (see Microcon Operating Instructions). Prepare the Microcon 100 concentrators by adding 100 µL of distilled deionized water to the filter side (top) of each concentrator.

13. Carefully remove the aqueous phase (supernatant of approximately 300 µL) from each sample, and transfer to the appropriate concentrator (avoid drawing any of the proteinaceous interface into the pipet tip).

14. Centrifuge the Microcon 100 concentrators for 5 min at $3000\times g$.

    Note: additional Centrifuge time may be required to filter the entire volume.

15. Discard the wash and return the filtrate cups to the concentrators.

16. Add 400 µL of filtered, distilled deionized water to the filter side of each Microcon 100 concentrator.

17. Centrifuge again at $3000\times g$ for 5 min and discard the filtrate cups.

18. Add 60 µL of hot (80°–90°C), filtered, distilled deionized water to the filter side of each Microcon 100 concentrator and place a retentate cup on the top of each concentrator.

19. Briefly vortex-mix the Microcon 100 concentrators with the retentate cups pointing upward.

20. Invert each concentrator with its retentate cup and centrifuge in a micro-centrifuge at $10\,000\times g$ for 3 min.

21. Discard the concentrators. Cap the retentate cups.

22. The samples can then be amplified. It is recommended that extracted DNA be amplified immediately or stored at 4°C until amplified.

Protocol for Using the Bessman Pulverizer When Bone Fragments are Too Small to be Cut and Sanded

Materials and Reagents

- Bessman Tissue Pulverizer
- Microcentrifuge tube
- Scalpel
- Stryker saw
- Dremel
- 1.5-mL Sarstedt tube

• SEB
• Proteinase K

Procedure

1. Prepare the reagent blank by swabbing the inside of the pulverizer with a sterile, dry cotton swab. A reagent blank should be prepared for each sample extracted utilizing the Bessman Tissue Pulverizer.

2. Cut off the cotton tip of the swab and place it in a sterile 1.5-mL microcentrifuge tube. Add 300 μL of SEB and 2 μL of 600 U/mL proteinase K to the tube.

3. Place the bone fragments into the pulverizer and strike the pestle with the hammer until the bone is soft and malleable.

4. Using a clean scalpel, cut the crushed bone into smaller pieces and place into a sterile 1.5-mL microcentrifuge tube to approximately half the volume of the tube.

5. Go to step No. 6 of bone extraction procedure.

Protocol for Using the SPEX 6700 Freezer/Mill When Accessible Marrow is Not Present

Procedure

1. Prepare the reagent blank by swabbing the inside of the sample container cylinder of the SPEX 6700 Freezer/Mill with a sterile, dry cotton swab. A reagent blank should be prepared for each sample container cylinder used.

2. Cut off the cotton tip of the swab and place it in a sterile 1.5-mL microcentrifuge tube. Add 300 μL of SEB and 2 μL of 600 U/mL proteinase K to the tube.

3. Cut the bone using a stryker saw or a dremel with a separating disk to a piece measuring approximately 2×5 mm. The separating disk should be used for only one sample and discarded.

4. Using a dremel with an emery disk, sand the outer surface of the bone so that the outer surface appears free of dirt and debris. The emery disk should be used for only one sample and discarded.

5. Place the sanded bone in the sample container cylinder of the Freezer/Mill. All cylinders should be thoroughly cleaned with a 10% bleach solution and exposed to UV light for a minimum of 15 min prior to reuse.

6. Keeping the top of the Freezer/Mill open, fill it with liquid nitrogen. Be careful not to expose face or hands to liquid nitrogen splashes. Approximately 5 L of liquid nitrogen are needed, most of which evaporates during the chill down period (about 10–15 min). Additional liquid nitrogen will be required to stabilize the liquid nitrogen level near the line inside the Freezer/Mill reservoir.

7. If a series of samples is to be ground, load at least three cylinders (each with an Impactor) and place one loaded cylinder into the grinding station (Solenoid bore) and the other two into the adjoining precooling stations.

8. Close the lid of the Freezer/Mill and wait until the stream of vapor decreases (about 4–5 minutes) while the samples in the cylinders are cooling.

9. Set the grinding time for approximately 3 min and adjust the impact frequency to the highest setting at which the impactor noise is a steady and uniform rattle. Do not open the lid of the Freezer/Mill during grinding. Should you wish to inspect the cylinder during the grinding cycle, turn the timer knob counterclockwise to the "Off" position before opening the lid.

10. When the grinding cycle is finished, open the lid, remove the cylinder with the Extractor and Cylinder Opener, and inspect the sample. If unground particles are visible, reinsert for additional grinding.

11. If the sample is sufficiently ground, place the cylinder on a flat surface and transfer one of the prechilled cylinders to the grinding position. Insert another loaded cylinder into the precooling station. Check the coolant level and replenish the liquid nitrogen as needed. Close the lid and start the grinding cycle.

12. Once the cylinder containing the sample has been removed from the Freezer/Mill, the cylinder may be left to warm to room temperature before removing the sample, or the cylinder may be opened at liquid nitrogen temperatures using the Extractor and Cylinder Opener. To open at liquid nitrogen temperatures, the cylinder is removed from the Freezer/Mill with the Extractor and placed into the Cylinder Opener. Holding the cylinder with the Extractor, place the other end of the cylinder in the hole and tighten the clamp knob. Swing the clamp section to its horizontal position. Place the cylinder section of the Extractor in its mating hole. Turn the knob clockwise until the end plug is pulled out from the cylinder.

Note: On rare occasions, liquid nitrogen may enter the interior of the cylinder during grinding. Then when the cylinder is exposed to room temperature, the expanding liquid nitrogen may forcibly "pop off" one or both end plugs of the cylinder. Therefore, while warming up the cylinder, place it in such a position that a possible pop off causes no injuries or damage.

13. Remove the impactor and empty the pulverized sample onto clean aluminum foil or wax paper. Carefully pour approximately 0.10 g pulverized sample into a sterile 1.5-mL microcentrifuge tube. A 1.5-mL Sarstedt tube filled approximately 1/3 yields the requisite amount of bone powder.

14. To this tube and the tube with the cotton swab tip (reagent blank) add 300 µL of SEB and 2 µL of 600 U/mL proteinase K.

15. Briefly vortex-mix and centrifuge the tubes in a microcentrifuge to force the bone fragments into the SEB.

16. Go to step No. 8 of the bone extraction procedure.

3.11 ALTERNATE EXTRACTION PROCEDURE FOR BONE

Materials and Reagents

- Razor blade
- Sandpaper
- 50-mL polypropylene tube
- Amicon Centricon® 30 microconcentrator tube (Millipore)

- 0.5 M EDTA, pH 7.5
- Saturated ammonium oxalate solution, pH 3.0
- Extraction buffer (10 mM Tris, 10 mM disodium EDTA-H$_2$0, 100 mM NaCl, and 39 mM DTT)
- Proteinase K (20 mg/mL)
- PCIA (25:24:1)
- Water-saturated *n*-butanol
- TE buffer

Procedure

1. Remove traces of soft tissue and marrow using a razor blade and sand-paper.

2. Crush bone into small fragments.

3. Grind the bone to a fine powder in the presence of liquid nitrogen in a metal blender. Transfer 5 g each of the powdered bone to three sterile 50-mL polypropylene tubes.

4. Add 40 mL of 0.5 M EDTA, pH 7.5, to each tube for decalcification. Gently agitate the tubes on a rotator at 4°C for 24 h.

5. Centrifuge the tubes at 2000× *g* for 15 min.

6. Discard the supernatant and add an additional 40 mL of 0.5 M EDTA, pH 7.5. Repeat the process daily over a three-to-five-day period.

7. Monitor the decalcification process by addition of saturated ammonium oxalate solution to the discarded supernatant. If the solution remains clear after the addition of ammonium oxalate, the decalcification process can be stopped.

8. Add 40 mL of sterile distilled, deionized water to each tube, agitate on a rotator for several min, and then centrifuge for 15 min at 2000× *g*. Discard the supernatant.

9. Repeat step No. 8 three times.

10. Add 2.0 mL of prewarmed (56°C) extraction buffer to the pellets in each tube and incubate at 56°C for 2 h with intermittent manual shaking of the tubes.

11. Allow extraction to continue for 10 h without agitation.

12. Add 100 µL of 20 mg/mL proteinase K solution to each tube and incubate at 56°C for 3 h with intermittent manual shaking.

13. Extract the solution in each tube three times with an equal volume of PCIA.

14. Combine the aqueous phase (i.e., the top layer) from the three tubes and extract one time in water-saturated *n*-butanol.

15. Concentrate the aqueous phase using a Centricon 30 microconcentrator tube.

16. Wash the retentate three times with 2 mL of TE buffer.

17. Store the DNA at 4°C.

3.12 TEETH

Before characterization of DNA from teeth samples, the teeth should be examined by a forensic dentist. All contributors should call the laboratory prior to submitting a case requiring analysis of DNA from teeth. The tooth or teeth should be photographed before the sample is extracted.

If several teeth are submitted for analysis, the order of preference for tooth selection for DNA extraction is as follows:

1. Nonrestored molar.

2. Nonrestored premolar.

3. Nonrestored canine.

4. Nonrestored anterior tooth.

5. Restored molar.

6. Restored premolar.

7. Restored canine.

8. Restored anterior tooth.

Protocol When Soft Tissue is Present

Materials and Reagents

- Vise (PanaVise®)
- Dremel
- Endodontic file
- SEB
- Proteinase K (600 U/mL)

Procedure

1. Prepare a reagent blank by adding 300 µL of SEB and 2 µL of 600 U/mL proteinase K to a sterile 1.5-mL microcentrifuge tube.

2. Place the tooth into a vise (PanaVise) and cut the tooth horizontally using a dremel with a separating disk at the cemento-enamel junction (demar-

cation on the tooth where the crown meets the root) being careful to avoid any restorations that may be present. The separating disk should be used for only one sample and discarded.

3. Remove the soft tissue with an endodontic file and small forceps, and place in a sterile plastic 1.5-mL microcentrifuge tube. The endodontic file should be used for only one sample and discarded.

4. Add 300 µL of SEB and 2 µL of 600 U/mL proteinase K.

5. Using a dremel with a small round bur, remove the dentin from the pulp chamber and place into a sterile 1.5-mL microcentrifuge tube. To this tube, add 300 µL of SEB and 2 µL of 600 U/mL proteinase K. The bur should be cleaned between each sample with alcohol swabs and a 10% bleach solution.

6. Briefly vortex-mix and centrifuge the tubes in a microcentrifuge to force tooth fragments or tissue into the SEB.

7. Go to step No. 8 of bone extraction procedure.

Protocol for Using the Bessman Pulverizer When No Soft Tissue is Present

Materials and Reagents

- Bessman Tissue Pulverizer
- Microcentrifuge tube
- Vise (PanaVise)
- Dremel

- SEB
- Proteinase K (600 U/mL)

Procedure

1. Prepare the reagent blank by swabbing the inside of the pulverizer with a sterile, dry cotton swab. A reagent blank should be prepared for each sample extracted utilizing the Bessman Tissue Pulverizer.

2. Cut off the cotton tip of the swab and place it in a sterile 1.5-mL microcentrifuge tube. To this tube add 300 µL of SEB and 2 µL of 600 U/mL proteinase K.

3. Place the tooth into a vise (PanaVise) and using the dremel with a separating disk, cut the tooth horizontally at the cemento-enamel junction (demarcation on the tooth where the crown meets the root) being careful to avoid any restorations that may be present. Make an additional horizontal cut in the tooth slightly apical to the original cut so that a small cross section of root can be separated from the remainder of the tooth. Discard the separating disk.

4. Using a dremel with an emery disk, sand the outer surface of the small cross section of tooth so that the outer surface appears free of dirt and

debris. The emery disk should be used for only one sample and discarded.

5. Place the sanded tooth section in the pulverizer and strike the pestle with the hammer until the tooth is soft and malleable.

6. Place the crushed tooth into a sterile, plastic 1.5-mL microcentrifuge tube to approximately 1/3 the volume of the tube.

7. To this tube add 300 µL of SEB and 2 µL of 600 U/mL proteinase K.

8. Briefly vortex-mix and centrifuge the tubes in a microcentrifuge to force tooth fragments or tissue into the SEB.

9. Go to step No. 8 of bone extraction procedure.

Protocol for Using the SPEX 6700 Freezer/Mill When No Soft Tissue is Present

Materials and Reagents

- SPEX 6700 Freezer/Mill
- Cotton swab
- Vise (PanaVise)
- Dremel
- Emery disk

- SEB
- Proteinase K
- Liquid nitrogen
- 10% bleach

Procedure

1. Prepare the reagent blank by swabbing the inside of the sample container cylinder of the SPEX 6700 Freezer/Mill with a sterile, dry cotton swab. A reagent blank should be prepared for each sample container cylinder used.

2. Cut off the tip of the swab and place it in a sterile 1.5-mL microcentrifuge tube. Add 300 µL of SEB and 2 µL of 600 U/mL proteinase K to the tube.

3. Place the tooth into a vise (PanaVise) and using the dremel with a separating disk, cut the tooth horizontally at the cemento-enamel junction (demarcation on the tooth where the crown meets the root) being careful to avoid any restorations that may be present. Make an additional horizontal cut in the tooth slightly apical to the original cut so that a small cross section of root can be separated from the remainder of the tooth. Discard the separating disk.

4. Using a dremel with an emery disk, sand the outer surface of the small cross section of tooth so that the outer surface appears free of dirt and debris. The emery disk should be used for only one sample and discarded.

5. Place the sanded tooth in the sample container cylinder of the Freezer/Mill.

6. Keeping the top of the Freezer/Mill open, fill it with liquid nitrogen. Be careful not to expose face or hands to liquid nitrogen splashes. Approximately 5 L of liquid nitrogen are needed, most of which evaporate during the chill down period, (about 10–15 minutes). Additional liquid nitrogen will be required to stabilize the liquid nitrogen level near the line inside the Freezer/Mill reservoir.

7. If a series of samples is to be ground, load at least three cylinders (each with an impactor) and place one loaded cylinder into the grinding station (Solenoid bore) and the other two into the adjoining precooling stations.

8. Close the lid of the Freezer/Mill and wait until the stream of vapor decreases (about 4–5 min) while the samples in the cylinders are cooling.

9. Set the grinding time for approximately 3 min and adjust the impact frequency to the highest setting at which the impactor noise is a steady and uniform rattle. Do not open the lid of the Freezer/Mill during grinding. Should you wish to inspect the cylinder during the grinding cycle, turn the timer knob counterclockwise to the "Off" position before opening the lid.

10. When the grinding cycle is finished, open the lid, remove the cylinder with the Extractor and Cylinder Opener, and inspect the sample. If unground particles are visible, reinsert for additional grinding.

11. If the sample is sufficiently ground, place the cylinder on a flat surface and transfer one of the prechilled cylinders to the grinding position. Insert another loaded cylinder into the precooling station. Check the coolant level and replenish the liquid nitrogen as needed. Close the lid and start the grinding cycle.

12. Once the cylinder containing the sample has been removed from the Freezer/Mill, the cylinder may be left to warm to room temperature before removing the sample, or the cylinder may be opened at liquid nitrogen temperatures using the Extractor and Cylinder Opener. To open at liquid nitrogen temperatures, the cylinder is removed from the Freezer/Mill with the Extractor and placed into the Cylinder Opener. Holding the cylinder with the Extractor, place the other end of the cylinder in the hole and tighten the clamp knob. Swing the clamp section to its horizontal position. Place the cylinder section of the Extractor in its mating hole. Turn the knob clockwise until the end plug is pulled out from the cylinder.

    Note: On rare occasions, liquid nitrogen may enter the interior of the cylinder during grinding. Then when the cylinder is exposed to room temperature, the expanding liquid nitrogen may forcibly pop off one or both end plugs of the cylinder. Therefore, while warming up the cylinder, place it in such a position that a possible pop off causes no injuries or damage. All cylinders should be thoroughly cleaned with a 10% bleach solution and exposed to UV light for a minimum of 15 min prior to reuse.

13. Remove the impactor and empty the pulverized sample onto clean aluminum foil or wax paper. Carefully pour 0.10 g of the pulverized sample into the sterile 1.5-mL microcentrifuge tube.

14. To this tube and the tube with the cotton swab tip (reagent blank) add 300 μL of SEB and 2 μL of 600 U/mL proteinase K.

15. Briefly vortex-mix and centrifuge the tubes in a microcentrifuge to force tooth fragments or tissue into the SEB.

16. Go to step No. 8 of bone extraction procedure.

3.13 CHELEX EXTRACTION OF DNA FROM WHOLE BLOOD OR BLOODSTAINS

Materials and Reagents

- 1.5-mL microcentrifuge tube
- Microcon 100 microconcentrator device
- Parafilm M™
- Phosphate buffered saline (PBS)
- 5% Chelex 100
- TE buffer

Procedure

1. Add 3 μL whole blood or a bloodstain approximately 3×3 mm to a sterile 1.5-mL microcentrifuge tube. Pipet 1 mL sterile PBS into the tube. Vortex-mix for 2 s.

2. Incubate at room temperature for 30 min. Vortex-mix for 5 s.

3. Centrifuge in a microcentrifuge for 1 min at maximum speed.

4. Without disturbing the pellet, carefully remove and discard the supernatant, leaving enough behind to cover the pellet without disturbing it. If the sample is a bloodstain, leave the fabric substrate in the tube with the pellet.

5. Add 200 μL 5% Chelex. Vortex-mix the tube for 10 s.

6. Incubate at 56°C for 30 min.

7. Vortex-mix the tube for 10 s.

8. Incubate in a boiling water bath for 8 min.

9. Vortex-mix the tube for 10 s.

10. Centrifuge in a microcentrifuge for 3 min at maximum speed. If no further purification is necessary proceed to step No. 15.

11. If removal of PCR inhibitors is required, then carefully transfer Chelex supernatant to the upper chamber of a Microcon 100 microconcentrator device. Avoid transferring any Chelex resin. If supernatant volume is less than 2 mL, bring volume to the 2 mL line on the Centricon device using TE.

12. Cover top of Centricon with a piece of parafilm. Poke a small hole in the

parafilm with a sterile pipet tip.

13. Centrifuge tubes at 1000× *g* for 20 min.

14. Invert upper chamber of Centricon into the conical collection tube. Centrifuge at 500× *g* for 5 min to collect retentate.

15. Store at 4°C or frozen. Prior to use of these samples for amplification, repeat step Nos. 9 and 10.

3.14 CHELEX EXTRACTION OF DNA FROM SEMEN-CONTAINING STAINS

Vaginal swabs should be air-dried and stored at -20°C.

Materials and Reagents

- 1.5-mL microcentrifuge tube
- Microcon 100 microconcentrator device
- Parafilm M

- PBS
- 20% Chelex 100
- 5% Chelex 100
- Sperm wash buffer
- TE buffer
- Deionized water
- Proteinase K (20 mg/mL)
- 1.0 M DTT

Procedure

1. Using a clean cutting surface for each different sample, dissect swab or fabric into thirds.

2. Add swab or fabric cutting to a sterile 1.5-mL microcentrifuge tube. Pipet 1 mL sterile PBS into the tube. Vortex-mix for 2 s.

3. Incubate at room temperature for 30 min.

4. Vortex-mix for 10 s.

5. Remove the swab or fabric using an applicator stick or pipet tip.

6. Centrifuge the sample in a microcentrifuge for 1 min at maximum speed.

7. Without disturbing the pellet, carefully remove and discard all but approximately 50 µL of the supernatant (or leave behind twice the volume of the pellet, if this would amount to more than 50 µL).

   Note: This pellet is called the cell pellet.

8. Add 150 µL sterile distilled, deionized water to the cell pellet. Add 1 µL of proteinase K (20 mg/mL). Vortex-mix for 1 s.

9. Incubate at 56°C for 1 h to lyse nonsperm cells.

10. Centrifuge in a microcentrifuge for 5 min at maximum speed. The resultant pellet is called the sperm pellet.

11. Add 150 μL of the supernatant to 50 μL of 20% Chelex in a fresh, sterile 1.5-mL microcentrifuge tube. This sample is called the female fraction and generally contains nonsperm DNA. Save for female fraction DNA analysis, which begins at step No. 16.

12. Wash the sperm pellet as follows: Resuspend the pellet in 0.5 mL sperm wash buffer. Vortex-mix briefly. Centrifuge in a microcentrifuge for 5 min at maximum speed. Remove approximately 450 μL of the supernatant and discard.

13. Repeat wash step No. 12 an additional 4 times, for a total of 5 washes.

14. Wash sperm pellet once with sterile distilled, deionized water as follows: Resuspend the pellet in 1 mL water. Vortex-mix briefly. Centrifuge in a microcentrifuge for 5 min at maximum speed. Remove and discard approximately 950 μL of the supernatant.

15. Add 150 μL of 5% Chelex to the approximately 50 μL sperm cell pellet (final volume should be approximately 200 μL). Add 1 μL of proteinase K and 7 μL 1.0 M DTT.

16. Vortex-mix female fractions and sperm samples for 10 s. Centrifuge in a microcentrifuge for 3 s.

17. Incubate at 56°C for 1 h.

18. Vortex-mix the tubes for 10 s.

19. Incubate the samples in a boiling water bath for 8 min.

20. Vortex-mix the tubes for 10 s.

21. Centrifuge in a microcentrifuge for 3 min.

22. Store samples at 4°C or frozen. Prior to use of these samples for amplification, repeat step Nos. 20 and 21.

3.15 CHELEX EXTRACTION OF DNA FROM SALIVA FROM ORAL SWABS, FILTER PAPER, OR GAUZE

Procedure

1. Using a clean cutting surface for each different sample, cut a 3 × 3 mm portion of the gauze or filter paper or dissect swab into thirds. Place in a sterile 1.5-mL microcentrifuge tube.

2. Proceed to step No. 5 of procedure entitled Chelex Extraction of DNA from Whole Blood or Bloodstains.

3.16 CHELEX EXTRACTION OF DNA FROM ENVELOPE FLAPS AND STAMPS

Procedure

1. Carefully open the envelope flap or remove the stamp using steam and clean tweezers.
2. Using a sterile cotton swab moistened in sterile distilled, deionized water, swab gummed envelope flap or stamp. Cut the cotton swab from the stick and transfer the cotton to a sterile 1.5-mL microcentrifuge tube.
3. Add 450 μL 5% Chelex to the swab.
4. Add 15 μL proteinase K (20 mg/mL) to each tube. Vortex-mix for 2 s.
5. Incubate at 56°C for 90 min.
6. Proceed to step No. 5 of procedure entitled Chelex Extraction of DNA from Whole Blood or Bloodstains.

3.17 CHELEX EXTRACTION OF DNA FROM CIGARETTE BUTTS

Procedure

1. Remove an approximately 5-mm–wide strip from the paper covering the cigarette butt in the area that would have been in contact with the mouth.
2. Pull paper slice from cigarette butt, cut into smaller pieces, and put pieces into a sterile 1.5-mL microcentrifuge tube.
3. Add 1 mL 5% Chelex to the tube.
4. Vortex-mix the tubes for 30 s.
5. Proceed to step No. 6 of procedure entitled Chelex Extraction of DNA from Whole Blood or Bloodstains.

3.18 DNA EXTRACTION FROM HAIR FOR mtDNA ANALYSIS

Materials and Reagents

- Micro Tissue Grinder
- Stratalinker® UV Crosslinker (Stratagene, La Jolla, CA, USA)
- Ultrasonic water bath

- Terg-a-zyme™ detergent
- 1.0 N H_2SO_4
- SEB
- Proteinase K (600U/mL)
- PCIA (25:24:1)
- Microcon 100 microconcentrator device
- Distilled, deionized water

Grinding the Hair

Micro Tissue Grinders are used to grind hairs for DNA extraction. The grinders consist of matched sets of mortars and pestles and should be used as such. To facilitate working with the grinders, the grinders may be placed in a small plastic tube, such as a 1.5-mL Sarstedt tube. Grinders should be used a maximum of three times and discarded. Prior to beginning hair extraction procedures, the grinders should be cleaned using the following protocol:

1. Carefully rinse the grinders with distilled deionized water. Using cotton tip applicators and warm 5% (wt/vol) Terg-a-zyme detergent, scrub the pestles and the insides of the mortars.

2. Rinse the grinders with distilled deionized water and add approximately 200 μL of 1.0 N H_2SO_4.

   Note: Gloves, protective eye wear, and a laboratory coat should be worn, and all work should be performed under a fume hood. Place the pestles in the matching mortars and briefly simulate grinding. Allow the mortars and pestles to soak in 1.0 N H_2SO_4 for a minimum of 20 min.

3. Rinse the grinders with distilled deionized water. Remove the pestles and place the mortars in a microcentrifuge. Pulse centrifuge at high speed to collect the remaining water. Remove the mortars from the microcentrifuge and remove the remaining water with a pipet.

4. Place the grinders in a rack and place in a Stratalinker for a minimum of 15 min.

DNA Extraction Procedure

1. Remove the hair to be extracted from the ethanol (or water) with clean forceps and examine it under a stereomicroscope for the possible presence of sheath material, surface debris, or bodily fluids.

2. Prepare a wash solution of 5% (wt/vol) Terg-a-zyme detergent.

3. Place the 2-cm portion of hair in a 1.5-mL plastic tube filled with approximately 1 mL of 5% Terg-a-zyme and place in a rack in an ultrasonic water bath. Agitate for approximately 20 min. Examine the hair for the presence of surface debris. If necessary, continue to wash the hair in the Terg-a-zyme solution until free of surface debris.

4. Briefly rinse the hair in 100% ethanol, followed by distilled deionized water.

5. A reagent blank should be prepared for each grinder to be used. Prepare the reagent blank by placing 200 μL of SEB into the micro tissue grinder to be used. Briefly simulate grinding. Remove the pestle and transfer the liquid to a sterile 1.5-mL plastic tube. Set aside until step No. 16.

6. To the same grinder add 200 μL of SEB. Place the hair fragment into the micro tissue grinder.

7. Move the pestle up and down to force the hair into the bottom of the mortar. Grind until fragments are no longer visible.

8. Remove the pestle from the mortar. If liquid is adhering to the pestle head, gently pass it along the inner lip of the mortar until liquid flows to the bottom of the mortar.

9. Transfer the homogenate liquid to a sterile 1.5-mL plastic tube.

10. Add 1 μL of 600 U/mL of proteinase K to each tube.

11. Vortex-mix on low speed and briefly centrifuge. Place the tubes in a water bath and incubate at 56°C for a minimum of 2 h and a maximum of 24 h.

12. Remove the tubes from the water bath and briefly centrifuge in a micro-centrifuge to force the condensate into the bottom of the tubes.

13. Add 200 μL of PCIA (25:24:1) to each tube.

   CAUTION: PCIA is an irritant and is toxic. Its use should be confined to a hood. Gloves and a mask should be worn.

14. Vortex-mix for 30 s to attain a milky emulsion, then centrifuge the tubes in a microcentrifuge for 3 min at $10\,000 \times g$.

15. Assemble and label a Microcon 100 unit for each sample (see Microcon Operating Instructions). Prepare the Microcon 100 concentrators by adding 200 μL of distilled deionized water to the filter side (top) of each concentrator.

16. Carefully remove the aqueous phase (supernatant of approximately 200 μL) of the PCIA from each tube and transfer to the appropriate concentrator. Avoid drawing any of the proteinaceous interface into the pipet tip.

17. Centrifuge the Microcon 100 concentrators for 5 min at $3000 \times g$.

   Note: Additional centrifuge time may be required to filter the entire volume.

18. Discard the wash and return the filtrate cups to the concentrators.

19. Add 400 μL of filtered, distilled deionized water to the filter side of each Microcon 100 concentrator.

20. Centrifuge again at $3000 \times g$ for 5 min and discard the filtrate cups.

21. Add 60 μL of hot (80°–90°C), filtered, distilled deionized water to the filter side of each Microcon 100 concentrator and place a retentate cup on the top of each concentrator.

22. Briefly vortex-mix the Microcon 100 concentrators with the retentate cups pointing upward.

23. Invert each concentrator with its retentate cup and centrifuge in a micro-centrifuge at $10\,000 \times g$ for 3 min.

24. Discard the concentrators. Cap the retentate cups.

25. Store samples at 4°C.

3.19 NONORGANIC EXTRACTION OF DNA FROM WHOLE BLOOD—SALTING OUT METHOD

Materials and Reagents

- Cell lysis buffer (0.32 M sucrose, 10 mM Tris-HCl, pH 7.6, 5 mM $MgCl_2$, and 1% Triton® X-100)
- Protein digestion buffer (10 mM Tris-HCl, pH 8.0, 10 mM NaCl, and 10 mM EDTA)
- Proteinase K (10 mg/mL)

Procedure

1. Whole blood specimens are collected in EDTA vacutainer tubes and stored at 4°C. Resuspend cells by manual mixing prior to extraction.
2. Place 0.5 mL of whole blood in 1.5-mL Sarstedt tube.
3. Add 1.0 mL of cell lysis buffer to the tube and vortex-mix briefly.
4. Centrifuge tube at 900× g (or 4000 rpm) for 5 min in a microcentrifuge.
5. Decant and discard the supernatant and blot the tube with absorbent paper.
6. Repeat steps 3 through 5.
7. Add 1.0 mL of protein digestion buffer to the tube and vortex-mix the tube to resuspend the pellet.
8. Centrifuge the tube at 900× g for 5 min in a microcentrifuge.
9. Decant and discard the supernatant and blot the tube with absorbent paper.
10. Place the tube with the pellet on wet ice.
11. Add 225 μL of protein digestion buffer and 25 μL of proteinase K solution to the tube. Pipet solution up and down to break up the pellet. Vortex-mix the tube for several s to resuspend the pellet.
12. Place the tube in a 65°C heat block and incubate for 2 h. Approximately every 20 min, vortex-mix the tube briefly to ensure the pellet is resuspended.
13. Remove the tube from the heat block and vortex-mix vigorously for 30 s.
14. Centrifuge the tube for 2 min at maximum speed in a microcentrifuge.
15. Store DNA sample at 4°C.

3.20 PREPARATION OF DNA IMMOBILIZED ON FTA PAPER

Blood deposited on FTA paper is an excellent source of genomic DNA. The paper facilitates handling, prevents degradation of the DNA, and enables storage of the material at ambient temperature. Cellular debris can be washed away, and the DNA remains immobilized in the FTA paper matrix. A punch of the purified DNA can be placed directly into a PCR.

Materials and Reagents

- Harris punch (1.2-mm diameter) (Fitzco)
- FTA purification reagent
- TE buffer (10 mM Tris-HCl, pH 8.0, 0.1 mM EDTA)

Procedure

1. Using a Harris punch, punch out a 1.2-mm diameter portion of the bloodstain and place it in a 0.5-mL PCR tube.

2. Add 200 μL of FTA purification reagent to the tube, cap the tube, and vortex-mix the tube briefly.

3. Incubate at ambient temperature for 5 min. Briefly vortex-mix the tube midway through the incubation.

4. Briefly vortex-mix. Remove and discard the supernatant.

5. Although not routinely necessary, steps 3 through 5 can be repeated up to a total of 3 washes.

6. Add 200 μL of TE buffer to the tube, cap the tube, and briefly vortex-mix the tube.

7. Incubate at ambient temperature for 5 min Briefly vortex-mix midway through the incubation.

8. Decant and discard the supernatant.

9. Allow the FTA punch to dry in the tube at ambient temperature (at least 1 h), or dry the punch at 60°C for 30 min.

10. The DNA-containing punch is ready for RFLP or PCR-based analysis.

SUGGESTED READING

1. Belgrader, P., S.A. Del Rio, K. Turner, M.A. Marino, K.R. Weaver, and P.E. Williams. 1995. Automated DNA purification and amplification from blood-stained cards using a robotic workstation. BioTechniques *19*:426-432.

2. Bever, R.A. and M.A. DeGuglielmo. 1994. Development and validation of buccal swab collection method for DNA testing. Adv. Forensic Haemogent. *5*:199-201.

3. Burgoyne, L. and C. Rogers. 1997. Bacterial typing: storage and processing of stabilized reference bacteria for polymerase chain reaction without preparing DNA—an example of an automatable procedure. Anal. Biochem. *247*:223-227.

4. Comey, C.T., B.W. Koons, K.W. Presley, J.B. Smerick, C.A. Sobieralski, D.M. Stanley, and F.S. Baechtel. 1994. DNA extraction strategies for amplified fragment length polymorphism analysis. J. Forensic Sci. *39*:1254-1269.

5. Del Rio, S., M.A. Marino, and P. Belgrader. 1996. Reusing the same bloodstained punch for sequential DNA amplifications and typing. BioTechniques *20*:970-974.

6. Gill, P., A.J. Jeffreys, and D.J. Werrett. 1985. Forensic application of DNA fingerprints. Nature *318*:577-579.

7. Gill, P., J.E. Lygo, S.J. Fowler, and D.J. Werrett. 1987. An evaluation of DNA fingerprinting for forensic purposes. Electrophoresis *8*:38-44.

8. Giusti, A.M., M. Baird, S. Pasquale, I. Balasz, and G. Glassberg. 1986. Application of deoxyribonucleic acid (DNA) polymorphisms to the analysis of DNA recovered from sperm. J. Forensic Sci. *31*:409-417.

9. Grimberg, J., S. Nawoschik, L. Belluscio, R. McKee, A. Turck, and A. Eisenberg. 1989. A simple and efficient nonorganic procedure for the isolation of genomic DNA from blood. Nucleic Acids Res. *17*:8390.

10. Hagelberg, E., B. Sykes, and R. Hedges. 1989. Ancient bone DNA amplified. Nature *342*:485.

11. Hochmeister, M.N., B. Budowle, U.V. Borer, U. Eggmann, C.T. Comey, and R. Dirnhofer. 1991. Typing of DNA extracted from compact bone from human remains. J. Forensic Sci. *36*:1649-1661.

12. Hochmeister, M.N., B. Budowle, J. Jung, U.V. Borer, C.T. Comey, and R. Dirnhofer. 1991. PCR-based typing of DNA extracted from cigarette butts. Int. J. Legal Med. *104*:229-233.

13. Jinks, D.C., M. Minter, D.A. Tarver, M. Vanderford, J.F. Hejtmancik, and E.R.B. McCabe. 1989. Molecular genetic diagnosis of sickle cell disease using dried blood specimens on blotters used for newborn screening. Hum. Genet. *81*:363-366.

14. Jung, J.M., C.T. Comey, D.B. Baer, and B. Budowle. 1991. Extraction strategy for obtaining DNA from bloodstains for PCR amplification and typing of the HLA-DQ alpha gene. Int. J. Legal Med. *104*:145-148.

15. Kanter, E., M. Baird, R. Shaler, and I. Balasz. 1986. Analysis of restriction fragment length polymorphisms in DNA recovered from dried bloodstains. J. Forensic Sci. *31*:403-408.

16. Maniatis, T., E.F. Fritsch, and J. Sambrook. 1982. Molecular Cloning: A Laboratory Manual. CSH Laboratory Press, Cold Spring Harbor, NY.

17. Rogan, P.K. and J. Salvo. 1990. Study of nucleic acids isolated from ancient remains. Yearbook Phys. Anthropol. *33*:195-214.

18. Sambrook, J., E.F. Fritsch, and T. Maniatis. 1989. Molecular Cloning: A Laboratory Manual, Vols. 1, 2, and 3. 2nd ed. CSH Laboratory Press, Cold Spring Harbor, NY.

19. Singer-Sam, J., R.L. Tanguay, and A. Riggs. 1989. Use of Chelex to improve the PCR signal from a small number of cells. Amplifications: A Forum for PCR Users *3*:11.

20. Walsh, S., D.A. Metzder, and R. Higuchi. 1991. Chelex 100 as a medium for simple extraction of DNA for PCR-based typing from forensic material. BioTechniques *10*:506-513.

Determination of the Quantity of DNA

4

Determining the quantity of DNA is desirable to enhance the quality of results when performing DNA typing, particularly for the polymerase chain reaction (PCR). Regulation of initial DNA template quantities for PCR can reduce undue consumption of biological evidence and reagents and can prevent potential preferential amplification of the larger alleles of variable number tandem repeat (VNTR) loci. The quantity of DNA typically is estimated by spectrophotometry or by analysis in ethidium bromide-stained agarose gels. Spectrophotometry measures the quantity of soluble DNA in a sample by absorbance at 260 nm. Yield gels are an alternate method to estimate both the quantity and quality of the DNA. The DNA is exposed to an intercalating dye, e.g., ethidium bromide, and then the sample is subjected to electrophoresis in an agarose gel. Under UV light, the ethidium bromide/DNA complex will fluoresce. To estimate the quantity of DNA, the intensity of fluorescence of the unknown sample is compared to that of control samples of known concentration. Generally, 1 to 5 ng of double-stranded DNA can be detected in ethidium bromide-stained yield gels. The sensitivity is not as effective for single-stranded DNA. Because the DNA is subjected to electrophoresis, a determination can be made of whether or not high molecular weight DNA has been recovered.

However, a quantitation method to detect subnanogram quantities of human-specific DNA is needed for some forensic biospecimens, particularly those that have been degraded substantially and may be commingled with bacteria and/or fungi. Ethidium bromide-stained agarose gels and spectrophotometry fail to distinguish between human and nonhuman genomic DNA, and thereby the usable DNA fraction for a human-specific assay may be overestimated. Additionally, single-stranded DNA that results from

DNA Typing Protocols: Molecular Biology and Forensic Analysis
By B. Budowle, J. Smith, T. Moretti, J. DiZinno
©2000 Eaton Publishing, Natick, MA

Chelex (Bio-Rad, Hercules, CA, USA) extraction or from environmentally induced damaged DNA can stain poorly with ethidium bromide; thus, the quantity of the DNA of interest may be underestimated. Moreover, the ethidium bromide-stained gel approach lacks sensitivity to detect less than 1–5 ng of DNA. Therefore, a sample that contains sufficient DNA for a PCR amplification may be below the threshold of detection of the quantitation assay.

A simple, human-specific method for the quantitation of genomic DNA has been developed and is based on a slot blot hybridization approach. Denatured, extracted DNA is immobilized on a nylon membrane and subsequently hybridized with the higher primate-specific alphoid probe to the D17Z1 locus. As a hybridization-based technique, this method is more effective (because of its insensitivity to contaminating nonhuman DNA) than the use of ethidium bromide-stained gels for estimating the appropriate quantity of human template DNA for PCR. The slot blot procedure can be completed within one working day, enables the simultaneous analysis of a large number of samples, can detect subnanogram quantities of human DNA, and has been designed to use either radioactive or nonradioactive reagents for detection. For the nonradioactive approach, chemiluminescent detection is employed that involves the light-emitting decomposition of LUMI-PHOS Plus® (Life Technologies, Gaithersburg, MD, USA). LUMI-PHOS Plus substrate yields a continuous light output for more than 48 hours, enabling chemiluminescent-based assays to fit conveniently into most laboratory protocols and routines.

4.1 QUANTIFICATION OF DNA OBTAINED FROM LIQUID BLOOD BY SPECTROPHOTOMETRY

The DNA obtained from a liquid blood sample is solubilized in 200 µL Tris-EDTA (TE) buffer. The following description of instrument operation is applicable specifically to the Beckman DU7 and appropriate modifications should be made for other spectrophotometers.

Materials and Reagents

- Model DU7 Spectrophotometer (Beckman Instruments, Fullerton, CA, USA)
- Microcuvette
- TE buffer

Procedure

1. The concentration of DNA in the sample will be determined using a microcuvette that enables the determination of absorbancy of samples with as little as 50 µL. This instrument remains on idle when not being used. To activate, push the ON button. Follow the menus to set the instrument to read dual wavelengths of 260 and 280 nm and to determine the 260/280 ratio.

2. The instrument will self-calibrate, and then request that the solvent blank be put in. Pipet 50 μl TE buffer into the microcuvette. Tap the cuvette gently on the laboratory bench to remove any air bubbles. Place the cuvette into the sample holder and push START. The instrument will determine the absorbancy at both 260 and 280 nm and store these readings.

3. Empty the cuvette. There is no need to rinse the cuvette this time. Place 50 μL of the DNA sample in the cuvette, put the cuvette in the instrument, and push RUN. Both 260 and 280 nm absorbancies will be measured and their ratio calculated. At the end of the session you can print out the entire list of values by pushing COPY.

4. Pipet the 50 μL of sample back into the original sample tube. Rinse the cuvette with water before the next sample is added to the cuvette.

5. DNA quantity calculations (an example):

Assume the readings are $A_{260} = 1.80$
$$A_{280} = 1.00$$
thus, the absorbancy ratio is $1.80/1.00 = 1.8$

The DNA content of the sample is calculated as follows:

$$(A_{260})(50^a)(0.2) = \mu g \text{ DNA}/200 \ \mu L$$
$$(1.80)(50)(0.2) = 18.0 \ \mu g \text{ DNA}$$
$$^a 50 \ \mu g \text{ DNA/mL yields an } A_{260} \text{ of } 1.0$$

4.2 ETHIDIUM BROMIDE/AGAROSE TEST GEL FOR ASSESSING THE QUALITY AND QUANTITY OF DNA ISOLATED FROM BODY FLUID STAINS—YIELD GEL

A number of precautions must be followed when preparing agarose gels. If liquid was lost due to evaporation when melting the agarose, water should be added to achieve the original volume after boiling. The agarose solution should be equilibrated to 55°C prior to casting to prevent distortion of both the gel and the plastic electrophoresis tray. If the gel is not submerged in buffer, samples are pipetted into the wells and topped-up with buffer where required—empty wells are filled with buffer. Minigels are run for 5 minutes, and large gels for 10 minutes at 40 V to facilitate movement of the samples into the gel. Buffer is then added to submerge the gel 2 to 3 mm. Alternatively, the buffer can be placed over the gel, and then the samples can be applied to the gel. Yield gels can be run under a variety of electrophoretic conditions.

Gels are stained with ethidium bromide, then destained (if needed), and photographed under UV light (302–316 nm) using a transilluminator and a Polaroid camera system.

An agarose gel used to estimate the quality and quantity of DNA recovered from forensic specimens requires certain controls:

1. Visual marker (λ DNA *Hind*III digest): indicates that the electrophoresis has proceeded properly as well as the approximate size of the DNA recovered, which can be useful for assessing if any DNA has degraded.

2. Yield gel calibration set (λ DNA): intact λ DNA (48502 bp) in quantities ranging from 300 to 10 ng. The DNA isolated from evidence items and reference samples are compared to these standards to semiquantitatively estimate the amount of DNA in the samples. They also represent high molecular weight DNA and, by comparison to the sample DNAs, can indicate the quality of the DNA recovered.

3. Human DNA control (K562 DNA): 200 ng of undigested K562 cell line DNA is included on the yield gel. This control sample provides a means to determine the precision of the estimated DNA quantities for the evidentiary DNA samples, as a known quantity of human DNA to be used for comparison to the λ DNAs in the calibration set.

Materials and Reagents

- Transilluminator
- Polaroid camera (Polaroid, Cambridge, MA, USA)
- H4 Tank (Figure 1; Life Technologies)
- Polaroid film (No. 553, Polaroid)

- *Hin*dIII digested λ DNA
- K562 cell line DNA
- TE buffer

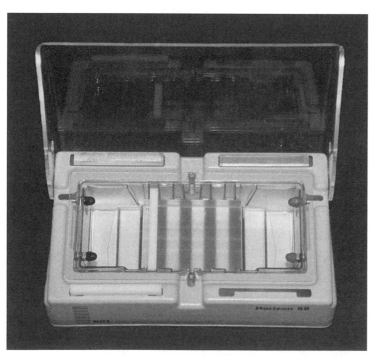

Figure 1. Horizontal submarine gel tank apparatus for agarose analytical gel electrophoresis (Life Technologies, Gaithersburg, MD, USA).

- Agarose gel loading solution
- Type II agarose (Sigma Chemical, St. Louis, MO, USA) or SeaKem® ME agarose (FMC BioProducts, Rockland, ME, USA)
- 1× Tris-acetate EDTA (TAE) buffer
- Ethidium bromide (EtBr)

Procedure

1. The DNA has been solubilized in 36 μL of TE buffer at 56°C for a minimum of 2 h. Vortex-mix briefly and centrifuge for 5 s. Note: You must centrifuge before the tube cap is opened.

2. Remove 4 μL DNA and combine with 2 μL loading solution. Centrifuge for 2 s.

3. Prepare yield gel as follows:

Setup of Agarose Template

Gel Size	Gel Vol (mL)	Wells/Origin	Origins
11 × 14 cm	100	14/16	2 rows

Use 1% agarose (Sigma type II or SeaKem ME) in 1× TAE buffer with EtBr at a ratio of 10 μL EtBr/100 mL TAE.

Gel Volume (mL)	g Agarose/Gel	μL EtBr
100	1.00	10.0

   a. Add 100 mL 1× TAE/EtBr to the agarose.

   b. Bring the solution to a boil to dissolve agarose.

   c. Allow the agarose solution to equilibrate at 56°C.

   d. Pour agarose into gel form (be sure the comb is in place)

   e. Let stand 15 min to gel.

4. Pour 900 mL 1× TAE/EtBr buffer into electrophoresis tank.

5. Place the gel into the tank. Enough buffer should be present to cover the gel. Remove comb.

6. Pipet the DNA sample mixed with loading solution (6 μL total volume) into the well in the submerged gel. Do not poke the pipet tip through the bottom of the gel.

DNA standards must be included on the gel. Quantitative DNA standards are available from commercial sources or can be made in-house. The volume of each standard loaded into a gel is a function of the concentration supplied by commercial vendors.

7. Well number 1 contains the 10 μL of *Hind*III-digested λ DNA.

8. Wells 2 through 8 contain the calibration DNA set in descending quantities.

9. Well 9 contains 200 ng of K562 DNA.

Note: If a double origin yield is used, the *Hind*III-digested λ DNA should be placed in the first well of the second origin. The calibration controls can be added to the second origin but are not required.

10. Set the voltage at 200 V. When the bromophenol blue tracking dye has moved 1–2 cm from the origin the run can be stopped. For 11 × 14 cm gels, this should be approximately 20 min.

11. Remove the gel from the tank. Examine the gel on a UV light transilluminator. Intact DNA will move as a band and will reside close to the origin. A smear from the origin to, or past the dye front indicates that the DNA has been fragmented and may not be suitable for restriction fragment-length polymorphism (RFLP) analysis. Take a photograph of the gel. Polaroid film No. 553, f5.6, 1 s, red filter in place.

CAUTION: Do not expose yourself to the UV light for an excessive amount of time. Always wear a full face shield when working with the transilluminator.

12. From the photograph, assess the quantity of DNA in the various samples by comparison with the DNA standards. Multiply your estimate by 8 to obtain the total quantity of DNA in the remaining 32 µL of sample.

SLOT BLOT HYBRIDIZATION DNA QUANTIFICATION USING ^{32}P-LABELED ALPHOID PROBE

Materials and Reagents

- The Convertible® Filtration Manifold System slot blot apparatus (Figure 2; Life Technologies)
- UV light box
- Intensifying screens

- 0.5 M NaOH-0.5 M NaCl
- Biodyne™ B nylon membrane (Pall, Port Washington, NY, USA)
- 2× standard saline citrate (SSC)
- Slot blot hybridization solution
- 0.2 M NaOH
- Slot blot wash solution
- Saran® or Glad® wrap
- Kodak XAR5 film (Eastman Kodak, Rochester, NY, USA)

Procedure

1. Wash slot blot apparatus in water prior to use.

2. Warm slot blot hybridization buffer and slot blot wash buffer to 65°C in an incubator.

3. Label a microcentrifuge tube for each sample to be quantified, including 8 tubes for standards. Add 200 µL 0.5 M NaOH-0.5 M NaCl to each tube.

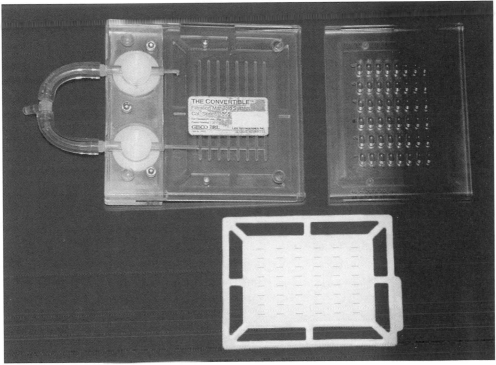

Figure 2. Slot blot apparatus; The Convertible Filtration Manifold System (kindly provided by Life Technologies).

4. Add approximately 1/10 of the DNA sample volume or a maximum of 20 µL to each tube for denaturation. Add standards to appropriate tubes: 20, 10, 4, 2, 1, 0.4, 0.2, and 0.1 ng. Incubate at room temperature for a minimum of 20 min.

5. Place nylon membrane (Pall Biodyne B) in a box containing 2× SSC. Soak for a minimum of 2 min.

6. Apply membrane to top of slot blot apparatus, turn on clamp vacuum, then turn on sample vacuum for 10 s.

7. Turn sample vacuum off.

8. Apply samples to wells in slot blot apparatus. When all samples have been added, turn on the sample vacuum very slightly (until air movement begins). Leave vacuum on for 5 min. If the sample has not drawn through completely, gently draw up fluid and reapply to the sample well to dislodge air bubbles. Avoid touching pipet tip to membrane.

9. Turn sample vacuum on to full and add 200 µL 0.5 M NaOH-0.5 M NaCl to each sample well. A repeater pipet can be used for this purpose. After addition to all sample wells, leave sample vacuum on for 2 more min.

10. Turn clamp and sample vacuums off, remove membrane, and place the membrane in a box containing slot blot membrane rinse buffer. Soak at room temperature for 5 min.

11. Remove membrane from box and lightly blot the membrane. Lay mem-

brane onto an UV light box with DNA side down. Expose to UV light for 90 seconds. Exposure times should be determined in-house by calibrating the UV light source. This can be accomplished by exposing controlled membranes (containing DNA) to various time intervals to determine the optimum exposure for the assay.

12. Thoroughly rinse effluent channel in base of slot blot apparatus using a squirt bottle filled with distilled water. This prevents a buildup of salt in the channel.

13. Place the membrane in a plastic box (e.g., approximately 22×22 cm or 9×16 cm). Up to eight membranes may be placed in a box. Add 57.9 mL prewarmed slot blot hybridization buffer (65°C). If more than one box is set up at the same time, incubate membranes in a shaking incubator at 65°C for 5 min to allow buffer to return to 65°C.

14. Denature probe, which is the human alphoid probe to D17Z1, by adding a volume of probe calculated to yield 5×10^5 disintegrations/minute/milliliter hybridization buffer to a tube containing herring sperm DNA (1.5 mL for 60 mL hybridization) and 0.2 M NaOH (0.6 mL for 60 mL hybridization). Place the tube in a boiling rack in a beaker of boiling water. Boil for 5 min.

15. Add denatured probe to hybridization buffer in container. Incubate in shaking water bath at 65°C for 1 h.

16. Pour off hybridization buffer into container designated for radioactive waste. Add enough 2× SSC to the box to cover the membranes and rinse for 10 to 20 s. Pour off 2× SSC into container designated for radioactive waste.

17. Add enough prewarmed slot blot wash buffer to the box to cover the membranes. Incubate in shaking incubator at 65°C for 15 min. Pour off wash into appropriate radioactive waste disposal facility.

18. Place membranes between two sheets of Whatman 1 blotting paper (Whatman, Clifton, NJ, USA) to remove excess solution. Wrap membranes in plastic wrap (e.g., Saran or Glad Wrap; use of other brands may cause marks on the autoradiogram). Take membranes to darkroom. Tape membranes, with DNA side down, to a sheet of Kodak XAR5 film (front film) and place a second sheet on the other side of the membrane (back film). Place films in cassette between 2 intensifying screens. Store at room temperature for short periods of time (less than an hour) or at -70° to -80°C for longer periods. (The process of autoradiography is described in Chapter 5 entitled Restriction Fragment-Length Polymorphism Typing of Variable Number Tandem Repeat Loci).

19. Develop the back film after 2 h. Develop the front film subsequently based on results seen on the back film. It may be desirable to obtain a range of exposures such that comparisons of standards with samples are made on exposures that are not overexposed.

20. Estimate the amount of DNA in each sample by comparison with appropriate standards. This represents the approximate amount of DNA in 20 μL of the original DNA sample. Calculate total DNA yield and DNA

concentration for each sample as follows:

yield: (total volume of extract)/(volume used for slot blot) × (nanograms in slot blot) = DNA yield (ng)

concentration: (total amount of DNA in sample)/(original volume of DNA sample) = DNA concentration

Example:

yield: (200 µL sample)/(20 µL for slot blot) × 10 ng = 100 ng total DNA in sample

concentration: 100 ng DNA/200 µL sample = 0.5 ng/µL

21. If an estimation of DNA yield cannot be accomplished due to the intensity excessively exceeding that of the standards, the sample may be diluted appropriately, and the slot blot procedure repeated.

22. Wash slot blot apparatus after use.

4.4 SLOT BLOT HYBRIDIZATION DNA QUANTITATION USING ALKALINE PHOSPHATASE (AP)-LABELED PROBE AND CHEMILUMINESCENCE DETECTION

Materials and Reagents

- The Convertible Filtration Manifold System slot blot apparatus (Life Technologies)
- UV light box
- Stratalinker® UV Crosslinker (Stratagene, La Jolla, CA, USA)

- Distilled water
- Slot blot hybridization solution
- Wash buffer
- Final wash solution
- Biodyne A nylon membrane (Pall)
- 2× SSC
- Alkaline phosphatase (AP)-labeled D17Z1 oligonucleotide probe
- LUMI-PHOS Plus
- 3 MM Whatman filter paper (Whatman)
- Plastic development folders
- Kodak XRP film (Eastman Kodak)

Procedure

1. Wash slot blot apparatus in distilled water prior to use.

2. Warm slot blot hybridization solution and wash buffer I to 50°C.

3. Label a microcentrifuge tube for each sample to be quantified. Add 200 µL of 0.5 M NaOH-0.5 M NaCl to each tube. Add sample DNA to each tube for denaturation; approximately 1/10 DNA sample volume. Incubate at room temperature for a minimum of 20 min.

4. Label a tube for each standard to be quantified. Add an appropriate volume of 0.5 M NaOH-0.5 M NaCl (200 μL per sample well; therefore, 400 μL are added for one membrane with 2 lanes of standards). Add an appropriate volume of standard DNA (10 μL per sample well; therefore, 20 μL are added for one membrane with 2 lanes of standards) to each tube for denaturation: 20, 10, 4, 2, 1, 0.4, 0.2, and 0.1 ng. Incubate at room temperature for a minimum of 20 min.

5. Place labeled nylon membrane (Pall Biodyne A) in a box containing 2× SSC. Soak for a minimum of 2 min.

6. Apply the membrane to top of slot blot apparatus, turn on clamp vacuum, then turn on sample vacuum for 10 s.

7. Turn sample vacuum off.

8. Apply samples to wells in slot blot apparatus. When all samples have been added, turn on the sample vacuum very slightly (until air movement is heard). Leave vacuum on for 5 min. If sample volume has not drawn through, gently draw up solution and reapply to sample well to dislodge air bubbles. Avoid touching pipet tip to membrane.

9. Turn sample vacuum on to full and add 200 μL 0.5 M NaOH-0.5 M NaCl to all sample wells and leave sample vacuum on for 2 more min.

10. Turn clamp and sample vacuums off, remove membrane, and place the membrane in a box containing slot blot membrane rinse buffer. Shake at room temperature for 5 min.

11. Remove membrane from box and drain. Lay membrane onto a UV light box or put in a Stratalinker for 90 s.

12. If multiple blots are being made, thoroughly rinse effluent channel in base of slot blot apparatus using a squirt bottle filled with distilled water. This prevents a buildup of salt in the channel.

    Note: Membranes can be maintained up to three days following transfer and prior to prehybridization by sandwiching between sheets of 3 MM Whatman filter paper and storing in plastic at room temperature.

13. Prepare 500 mL of 2× wash buffer and 500 mL of 1× final wash. Keep 2× wash buffer at 50°C and 1× final wash at room temperature.

14. Prehybridize membranes: incubate membranes in 30 mL hybridization solution at 50°C for 20 min in shaking water bath. Do not decant this solution.

15. Add 7.5 μL probe to box containing membranes and the 30 mL of hybridization solution used during the prehybridization. Incubate at 50°C for 20 min in a shaking water bath.

16. Decant hybridization solution. Add 250 mL of 2× wash buffer and shake at 50°C for 10 min. Decant and repeat.

17. Decant 2× wash buffer. Add 250 mL of 1× final wash buffer and shake at room temperature for 5 min. Decant and repeat.

18. Decant 1× final wash buffer. Blot membranes on Whatman paper and remove excess liquid. Do not let membrane dry out.

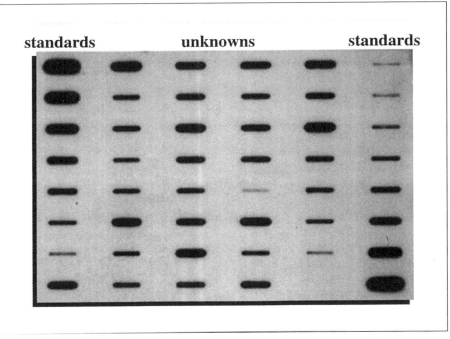

standards **unknowns** **standards**

Figure 3. Slot blot quantitation results developed using chemiluminescence. The standards are known amounts of human DNA that are used to determine (visually) the approximate amount of DNA in the unknown samples.

19. Place the membrane(s) into a box containing Lumi-Phos Plus for 5 min with gentle rocking. Approximately 1 to 4 membranes can be soaked in 15 mL of Lumi-Phos Plus (or 5–8 in 25 mL).

20. Place membranes in development folders.

21. Heat-seal the folders and trim. Wipe away any Lumi-Phos Plus that might have collected on the outside of the folder with a water-dampened laboratory wipe. Antistatic spray can be applied, if necessary.

22. Load into cassette with film. Store cassette at room temperature.

23. Remove the film after 30 min to 2 h. If additional exposure is needed, reload the cassette with a new film and the membrane. Overnight ramped membranes only require 5 to 10 min exposure times.

24. Estimate the amount of DNA in each sample by comparison with the appropriate standards (Figure 3). Calculate the total DNA yield and DNA concentration for each sample. A proportional decreasing level of intensity should be observed for each standard proceeding from 20 through 0.1 ng.

25. If an estimation of DNA yield cannot be accomplished due to sample intensity excessively exceeding that of the standards, the sample should be diluted appropriately, and the slot blot procedure repeated.

SUGGESTED READING

1. Budowle, B., F.S. Baechtel, C.T. Comey, A.M. Giusti, and L. Klevan. 1995. Simple protocols for typing forensic biological evidence: chemiluminescent detection for human DNA quantitation and RFLP analyses and manual typing of PCR amplified polymorphisms. Electrophoresis *16*:1559-1567.

2. Maniatis, T., E.F. Fritsch, and J. Sambrook. 1982. Molecular Cloning: A Laboratory Manual. CSH Laboratory Press, Cold Spring Harbor, NY.

3. Sambrook, J., E.F. Fritsch, and T. Maniatis. 1989. Molecular Cloning: A Laboratory Manual, Vols. 1, 2, and 3. 2nd ed. CSH Laboratory Press, Cold-Spring Harbor, NY.

4. Walsh, P.S., J. Varlaro, and R. Reynolds. 1992. A rapid chemiluminescent method for quantitation of human DNA. Nucleic Acids Res. *20*:5061-5065.

5. Waye, J., L. Presley, B. Budowle, G.G. Shutler, and R.M. Fourney. 1989. A simple and sensitive method for quantifying human genomic DNA in forensic specimen extracts. BioTechniques *7*:852-855.

Restriction Fragment-Length Polymorphism Typing of Variable Number Tandem Repeat Loci

5

Restriction fragment-length polymorphism (RFLP) analysis was the first DNA typing method used in human identity testing and generates DNA fragments of different length by restriction endonucleolytic digestion. The RFLP approach entails: *(i)* extraction of DNA; *(ii)* digestion of the DNA into fragments using a specific restriction endonuclease; *(iii)* electrophoretic separation of the fragments, based on size, by agarose gel electrophoresis; *(iv)* denaturing the double-stranded DNA fragments in a high pH environment; *(v)* transferring the single-stranded molecules out of the gel by capillary action onto a membrane support; *(vi)* hybridizing the immobilized DNA fragments with specifically labeled DNA probes; and *(vii)* detection of the hybrid products by autoradiography or chemiluminescence.

Originally, RFLP analysis was used to detect the presence or absence of specific, short DNA sequences called restriction sites. A restriction enzyme recognizes this short sequence along the double-stranded DNA and cuts the DNA wherever the specific site resides. For example, the restriction enzyme *Hae*III recognizes and cuts the DNA at the sequence GGCC (Figure 1).

Thus, the DNA from an individual is cut into many fragments. Because of sequence differences (i.e., in the enzyme recognition sequence among individuals), individuals may have restriction fragments of different lengths that can be used for comparisons (Figure 2).

At first glance, genomic DNA might not seem to be a likely source for individual characterization or identification by RFLP analysis because most of the coding regions of DNA among humans is similar. However, there are genetic polymorphisms that exist in the human genome that do not encode proteins and are highly polymorphic. One class of these genetic markers is known as variable number tandem repeats (VNTRs) or minisatellites. The VNTRs are comprised of tandemly repeated sequences (usually 9–80 bases in length per repeat unit) that exhibit variation in the number of repeats for alleles within and among individuals. Following digestion with a restriction enzyme, the length of each fragment is determined by the number of repeats

DNA Typing Protocols: Molecular Biology and Forensic Analysis
By B. Budowle, J. Smith, T. Moretti, J. DiZinno
©2000 Eaton Publishing, Natick, MA

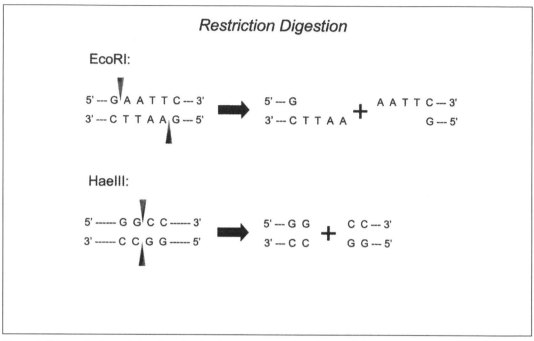

Figure 1. Schematic of restriction digestion for the restriction endonucleases *Eco*RI and *Hae*III. *Eco*RI is a six base cutter that recognizes the sequence **GAATTC**; the digestion results in sticky ends (staggered) at the site. *Hae*III is a four base cutter that recognizes the sequence **GGCC**; the digestion results in a blunt end cut.

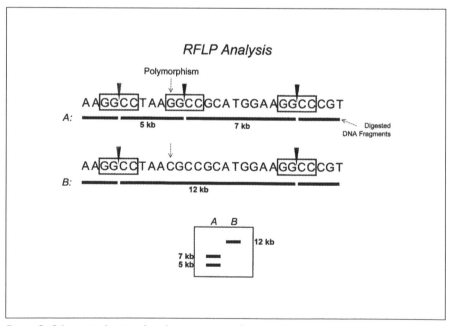

Figure 2. Schematic showing that the presence or absence of a restriction site can impact on the size of RFLP fragments. The restriction site **GGCC**, which is recognized by the enzyme *Hae*III, is present in three places in strand **A** and only two places in strand **B**. Therefore, two fragments (lengths of 5 and 7 kb) are generated in **A** and only a single 12-kb fragment is generated in **B**.

contained within each fragment. Many VNTR loci used for human identity testing exhibit more than 100 types in a population. In fact, such a high degree of polymorphism is exhibited that the typing of five to eight markers is sufficient to differentiate most, if not all, unrelated individuals. In other words, a multiple locus VNTR profile is extremely rare. More importantly, typing VNTR loci currently provides the scientist the best avenue to exclude a suspect who has been falsely associated with an evidentiary sample. In addition, typing can be accomplished, at times, with less than 50 ng of high molecular weight genomic DNA.

One factor that affects the effectiveness of RFLP analysis is the availability of well-characterized VNTR loci. The VNTR loci must be compatible with the restriction enzyme utilized for RFLP analysis (for the FBI laboratory that would be *Hae*III). Compatibility refers to the repeat sequence of the VNTR, which usually does not contain the restriction site specific to the restriction enzyme used in the assay. The loci alleles should generally fall in a size range that is greater than 500 bp and less than 20 000 bp. The loci routinely typed are D1S7, D2S44, D4S139, D5S110, D10S28, and D17S79 (Table 1). Additionally, VNTR loci are highly polymorphic and have a high degree of sensitivity of detection.

For human identification, DNA typing using RFLP analysis has been applied predominately to characterization of body fluid stains found at crime scenes and to resolving paternity and immigration cases. Figure 3 illustrates a DNA profile comparison of a mother, child, and alleged father(s) for a routine paternity case. Figure 4 shows a schematic of an RFLP analysis from a rape case comparing DNA profiles from a victim and an alleged rapist.

Table 1. Characteristics of Commonly Used VNTR Loci that are Compatible with *Hae*III

Locus	Probe	Chromosome Location	K562 Fragment Size in base pairs[a]
D1S7	MS1	1p33-35	4585, 4237
D2S44	YNH24	2q21.3-q22	2905, 1788
D4S139	pH30	4q35	6474, 3440
D5S110	LH1	5	3700, 2926
D10S28	TBQ7	10pter-p15	1754, 1180
D17S79	V1	17	1979, 1514

[a]Mean values from Standard Reference Material 2390 from National Institute of Standards and Technology.

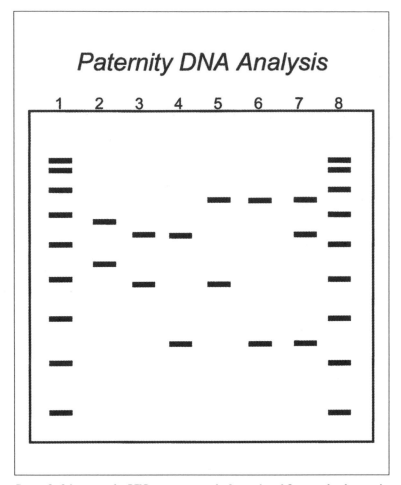

Figure 3. Schematic of a RFLP paternity result. Lanes 1 and 8 are molecular weight markers; lane 2 is the control sample profile; lane 3 is the mother's profile; lane 4 is the child's profile; lane 5 and 6 are different alleged fathers' profiles; and lane 7 is a mixed sample profile of the child and father in lane 6. The father in lane 5 is excluded, but the father in lane 6 cannot be excluded based on the analysis.

ELECTROPHORESIS

DNA molecules, regardless of size, have the same charge-to-mass ratio. Thus, all DNA fragments separated based on charge will migrate at the same rate and cannot be resolved. Therefore, digested double-stranded DNA fragments are separated based on size by electrophoresis through a sieving medium, and the electrophoretic system is performed using submarine gels. The medium typically used for RFLP typing is agarose (1%). The horizontal, agarose gels are submerged beneath buffer to maintain phase continuity and to enable effective heat dissipation in the relatively thick gels (5–10 mm) (Figure 5). Generally, fragments from 500 to 25 000 bp in length can be separated.

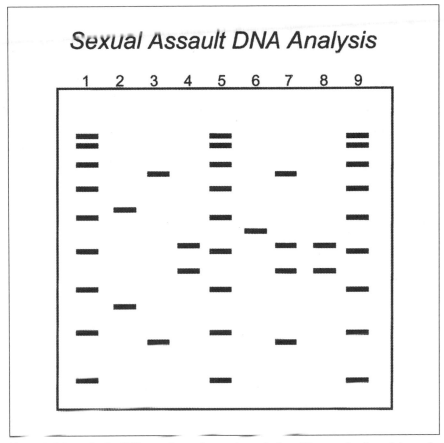

Figure 4. Schematic of a RFLP sexual assault case. Lanes 1, 5, and 9 are molecular weight markers; lane 2 is the control sample profile; lane 3 is the victim's profile; and lane 4 is the suspect's profile. Note that the reference samples are separated from the evidence profiles by a molecular weight marker (lane 5). Lane 6 is a profile from a bloodstain. Lane 7 is a mixed sample profile from the female fraction of a differential extraction. Lane 8 is a male fraction profile from the same differential extraction that the sample in lane 7 originated. The suspect in lane 4 is excluded as the source of the bloodstain sample in lane 6, but cannot be excluded as the source of the DNA in lane 8.

SOUTHERN BLOTTING

Southern blotting is the transfer of the electrophoretically-separated array of digested DNA fragments out of the gel and onto a membrane support (either nitrocellulose or nylon). The blotting relies on a flow, by capillary action, of a transfer solution from a reservoir through the agarose gel to a membrane overlaid by a stack of dry paper towels or blot pads. The DNA fragments are carried along with the flow of transfer solution from the gel to the membrane. Under appropriate conditions, the DNA readily binds to the membrane, maintaining the same array as it had at the end of electrophoresis. At some point before reaching the membrane, the DNA fragments must

Figure 5. Horizontal submarine gel tank apparatus for agarose analytical gel electrophoresis (H5; Life Technologies).

be denatured to single-stranded DNA so that the probe may bind during hybridization.

Two protocols for blotting are described—an alkali transfer to a positively charged nylon membrane, which is compatible with autoradiographic detection; and a high salt transfer to a neutrally charged nylon membrane, which is compatible with chemiluminescent detection. A low ionic strength, alkaline environment, which enables covalent binding of DNA to charged nylon membranes, is simple to make (i.e., 0.4 M NaOH) and also denatures the DNA during transfer. In contrast, a high salt transfer system first requires a denaturation of the DNA step and then a neutralization step of the gel prior to setting up the transfer.

There are several configurations of the basic theme of Southern blotting analysis. However, the simplest design is shown in Figure 6. It does not require any specialized equipment, and transfer is complete within six hours.

MEMBRANES

Nitrocellulose can bind DNA efficiently, but it is fragile (becoming brittle and falling apart). Nylon is more pliable and easier to handle. Currently, nylon membrane is the filter of choice. Efficient DNA binding is desirable so that the target DNA will not leach off the membrane after usage. UV fixing with neutral-charged membranes or basic pH and positive-charged membranes have been used to effectively immobilize DNA to nylon. The DNA should be single stranded when bound.

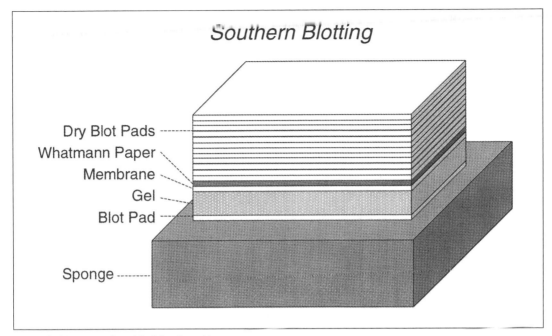

Figure 6. Schematic of modified Southern blot apparatus.

PROBE

Any fragment of nucleic acid can be used as a hybridization probe as long as it can be labeled so that the duplex can be detected (Figure 7).

The choice of probe (or probe design) depends on the typing technology, the availability of the probe, and the degree it can be labeled. DNA can be cloned into plasmids or bacteriophages. Thus, probe yield can be increased, and stability can be maintained. The vector should not contain sequences

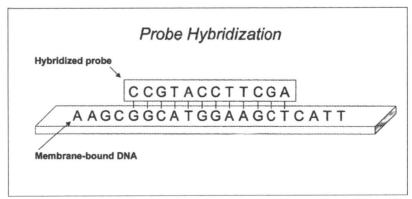

Figure 7. A probe is a short piece of single-stranded DNA that is labeled with a tag for subsequent detection. In RFLP typing, under appropriate analytical conditions, the probe will hybridize to its complement that is immobilized on the nylon membrane.

that cross-react with the target sequences of the probe. Otherwise, the vector sequences may have to be removed prior to using the probe. The use of double-stranded probes encounters two competing reactions, which are reassociation of the probe and hybridization to the immobilized DNA. Hybridization with single-stranded probes does not have to address reassociation with the probe's complement. Synthetic probes offer an alternative in that an enzyme or other molecule (e.g., biotin) can be coupled directly to the probe. The longer the probe, the greater the specificity, but hybridization times are longer than that for shorter probes.

PROBE LABELING

Probes are labeled either isotopically or nonisotopically. ^{32}P is the most commonly used radioisotope. Detection is sensitive, but may take long times to develop (depending on the amount of target). Half-life is 14.3 days—so labeled probes have a short shelf-life. Radioactive probes can be labeled with ^{32}P to a specific activity greater than 10^5 counts per minute (cpm)/µL using commercially available labeling kits. Typically 50- to 100-ng aliquots of probe are labeled. Prior to hybridization, the probe is denatured by boiling for 5 minutes followed by quenching on ice. The process of nick translation utilizes DNase I to create single-strand nicks in double-stranded DNA. The $5' \rightarrow 3'$ exonuclease and $5' \rightarrow 3'$ polymerase actions of *Escherichia coli* DNA Polymerase I are then used to remove stretches of single-stranded DNA starting at the nicks and replace them with new strands made by the incorporation of labeled deoxyribonucleotides. As a result, each nick moves along the DNA strand and is repaired in a $5' \rightarrow 3'$ direction. Nick translation can utilize any dNTP labeled with ^{32}P.

Nonradioactive labeling can allow for the incorporation of biotinylated nucleotides into DNA by standard techniques, such as nick translation or by direct labeling. Alternatively, an enzyme can be covalently linked to the probe directly or bound indirectly. These assays present less biohazards. The probes are far more stable than radioactively labeled probes and do not have the half-life issues related to radioactivity. Detection is rapid, and better quality control is attained. High probe concentrations can be used to drive the hybridization assay. Reagents are relatively inexpensive. Alkaline phosphatase-labeled oligonucleotide probes for VNTR loci and molecular weight markers are commercially available.

HYBRIDIZATION

Membrane hybridization was central to developing RFLP and initial polymerase chain reaction (PCR)-based assays. Hybridization is the annealing of a complementary probe to membrane-immobilized genomic target DNA (or vice versa). Basically, for RFLP typing, denatured DNA is immobilized on an inert support, such as nitrocellulose or nylon, so that it is accessible to incoming single-stranded probes. The probes are labeled to facilitate detection of the probe–target duplex.

The hybridization solution for probing VNTR sequences immobilized to nylon membranes has been considerably simplified relative to traditional formulations containing formamide, Denhardt's solution, dextran sulfate, or other additives. Only sodium dodecyl sulfate (SDS), polyethylene glycol (PEG), and phosphate buffer are used, and a prehybridization step is eliminated because of the addition of SDS at high concentration (7%) essentially blocks non-DNA bound sites on the membrane.

To distinguish between similar related sequences, reaction conditions should be optimized for the application. Factors that affect hybridization rates are: length of the fragments, base composition, ionic strength (cations; stringency), viscosity, denaturing agents (used to reduce the hybridization temperature because of fragility of nitrocellulose membranes), and temperature (stringency). Single-stranded probes are favored over denatured probes because re-annealing is avoided. High probe concentration drives the reaction, but too high a concentration should be avoided as it will lead to nonspecific hybridization. The rate of hybridization is decreased with increasing length of probe. The rate increases with GC content, but the effect usually is not substantial. Temperature affects hybridization rate, which is slow at low temperatures and increases to a broad range usually 20° to 25°C below the desired melting temperature (T_m) for annealing. At high temperatures, the strands tend to dissociate. The use of formamide decreases the T_m and has been used to reduce the hybridization temperature to 35° to 45°C. At low ionic strength (low salt), DNA fragments hybridize very slowly. High salt environments tend to stabilize mismatched duplexes. Dextran sulphate can be used to increase the hybridization rate (10%–tenfold) due to exclusion of the DNA from the volume occupied by the polymer—effectively increasing the DNA concentration (probe) or by inducing probe concatenation. However, PEG is preferred over dextran sulfate, primarily because it is easier to solubilize.

Hybridization generally is carried out in plastic sandwich boxes or in roller bottles. The membranes should be completely wetted and submerged in the hybridization solution. Large air bubbles trapped next to the membrane should be avoided, as these bubbles will impede probe hybridization. Gentle shaking should occur during the process.

POST-HYBRIDIZATION WASHES

Post-hybridization washes are carried out to remove loosely bound probe that could lead to nonspecific membrane background staining. Wash stringency increases as the solution temperature is increased and the buffer salt concentration is decreased. As the wash stringency increases, greater amounts of mismatched probe are removed from target DNA.

AUTORADIOGRAPHY

For DNA typing of single-copy genomic targets by RFLP, sensitivity of detection requirements often dictated that ^{32}P-labeled probes be utilized. The detection of the isotopic label is facilitated by autoradiography using high-speed X-ray film. The radioactive object (generally on a membrane) normally is placed in contact with X-ray film, and the energy released from the decay products of the radioisotope is absorbed by silver halide grains in the film emulsion to form a latent image. A chemical development process amplifies the latent image and renders the image visible on the film. Because the majority of emissions from ^{32}P pass through the thin film emulsion without contributing to the final image, the detection process can suffer from long exposure times and lack of sensitivity. Therefore, the membrane is sandwiched between X-ray film, and this complex is sandwiched between intensifying screens and exposed at -70°C (Figure 8).

Intensifying screens are required to convert the high energy radiation that passes through the film to emitted light, which exposes the film in the same spatial pattern as the emissions from the radioactively labeled material. However, sometimes autoradiographic exposures in excess of five days, even with intensifying screens, may be necessary to detect small quantities of a single-copy genomic target. The low temperature is used to overcome low intensity reciprocity failure of film. Another consideration is that X-ray film has a limited dynamic range of only two orders of magnitude. This makes comparisons of very different quantities of DNA on the same film exposure problematic and limits the ability to quantify images over a wide quantity range by optical density measurements.

CHEMILUMINESCENCE

Some scientists and/or laboratories have been reluctant to implement the RFLP method because of the handling and safety concerns that accompany the use of radioactively labeled probes. An alternative to the use of radioactively labeled probes is an approach that covalently links alkaline phosphatase directly to DNA probes. The annealed probe–target hybrid can be detected using a variety of reagents, particularly chemiluminescence substrates. The chemiluminescent-based detection system has supplanted ^{32}P labeling for RFLP typing, as well as for slot blot hybridization DNA quantitation (see Chapter 3 entitled Determination of the Quantity of DNA).

Application of chemiluminescent detection to RFLP typing requires a system with continuous light output so that signal can be collected over time (for increased sensitivity) and, if desired, so that multiple exposures to film can be made. The most sensitive chemiluminescent systems are those that emit a continuous glow. These systems have been applied widely to genetic research and involve the selective cleavage of stabilized 1,2-dioxetanes. One particularly useful substrate is Lumi-Phos Plus® (Life Technologies, Gaithersburg, MD, USA). The Lumi-Phos Plus substrate yields a continuous light output for more than 48 hours. The continued output enables chemiluminescent-based assays to fit conveniently into any laboratory protocol and routine, because film recording need not be performed immediately.

Chemiluminescence offers several advantages compared with autoradiography. These include: increased stability of the probes, enhanced resolution of hybridized fragments, film exposure at ambient temperature, faster detection, and more effective quality control of the probes (Figure 9).

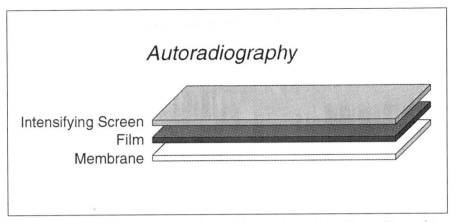

Figure 8. Schematic of orientation of DNA-containing membrane, X-ray film, and intensifying screen. An X-ray film and an intensifying screen can be placed under the membrane as well, so that two images may be produced.

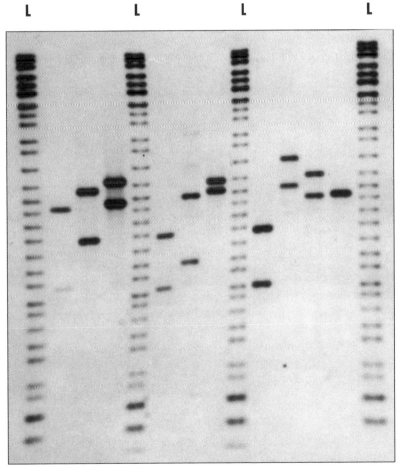

Figure 9. A lumigraph (30 minute exposure to film) of the VNTR locus D2S44. L = molecular weight ladder.

Film emulsions and hence exposure times may vary from lot to lot and between manufacturers. Thus, exposure times for each lot of film may have to be determined empirically. Because of the rapid detection of the chemiluminescent assay, film testing is needed more so than with autoradiography.

STRIPPING

Reprobing immobilized DNA with a series of different probes can extract more information from a single preparation, which is useful when DNA is limited or for throughput in high volume laboratories to reduce labor. To identify other genomic sites, the currently bound probe is removed by a process known as stripping. Stripping off the probe can be accomplished by exposure to high temperature or less harsh temperatures if a denaturant such as formamide is used. Do not let the filters dry out between strippings/probings or probes may irreversibly bind. Failure to remove a probe is not too much of a problem for southern blots, unless the same site is to be probed subsequently. The signal of the bound probe can be allowed to decay below detection levels, and then another probe may be hybridized to the immobilized DNA.

Figure 10. The restriction-digested DNA can be evaluated in the presence of ethidium bromide (DNA intercalator) and under exposure to UV light. The samples are *Hae*III-digested DNA.

5.1 DIGESTION AND REPRECIPITATION OF DNA ISOLATED FROM LIQUID BLOOD

Digestion of DNA with Restriction Endonuclease *Hae*III

Materials and Reagents

- Speed-Vac® Centrifuge (Savant Instruments, Holbrook, NY, USA)
- *Hae*III
- Water
- Ammonium acetate
- Absolute ethanol
- 70% ethanol
- Tris-EDTA (TE) buffer

Procedure

1. Combine the following:

$$200 \ \mu L \quad DNA$$
$$25 \ \mu L \quad \text{restriction buffer concentrate (REact 2)}$$
$$x \ \mu L \quad Hae III$$
$$\underline{y \ \mu L} \quad \text{water}$$
$$250 \ \mu L$$

Where: $x = (5)(\mu g \ DNA)/(U \ HaeIII/\mu L)$
                    $y = 25 - x$

2. Incubate at 37°C overnight.

Note: The concentration of restriction enzyme and incubation time may be increased to ensure that samples are completely digested. Caution should be exercised to monitor potential star activity with an exceedingly high concentration of restriction enzyme, and excessive levels of glycerol may inhibit the digestion.

Reprecipitation of *Hae*III-Digested Liquid Blood DNA

Procedure

1. To the 250 μL of restriction digest, add 83 μL 7.0 M NH_4OAc and mix.
2. Add 666 μL 100% EtOH and mix by hand.
3. Place tube at -20°C for 15–30 min. Do not let the tube freeze.
4. Centrifuge tube for 15 min. Decant and discard the alcohol.
5. Rinse pellet with 1.0 mL room temperature 70% EtOH. Centrifuge 5 min and decant and discard supernatant fluid. Remove remaining EtOH with a micropipet.

6. Put tube in Speed-Vac to remove residual alcohol, which takes approximately 15–30 min.

7. Add 16 µL TE buffer to the tube and place at 56°C to solubilize the DNA. After restriction, the DNA should solubilize within 30–60 min.

8. The DNA is now ready for another mini-test gel (this one to assess the completeness of restriction) and then an analytical gel.

Smaller Volumes of DNA, such as DNA Recovered from a Stained Material, can be Digested as Follows:

Procedure

1. Combine the following:

    32 µL DNA
     4 µL restriction buffer concentrate (REact 2)
     <u>4 µL</u> *Hae*III (40 international U)
    40 µL

    Note: Keep *Hae*III on ice.

    Mix by hand and centrifuge for 2 s.

    Note: The volume of restriction enzyme added should never be more than 10% of the final digestion volume. Also, keep the restriction enzyme on ice.

2. Incubate at 37°C overnight.

Reprecipitation of Digested DNA

Procedure

1. Centrifuge tubes for 5 s. To the 40 µL of DNA digest, add 13 µL of 7.0 M ammonium acetate and mix by hand.

2. Add 106 µL cold absolute EtOH and mix by hand.

3. Place tube at -20°C for 15–30 min. Do not let the tube freeze.

4. Centrifuge tube 15 min. Decant and discard the alcohol.

5. Rinse pellet with 1.0 mL room temperature 70% EtOH. Centrifuge for 5 min and decant and discard supernatant fluid. Remove remaining EtOH with a micropipet.

6. Put the tube in a Speed-Vac to remove residual alcohol, which takes approximately 15–30 min.

7. Add 16 µL TE buffer to the tube and place at 56°C to solubilize the DNA. After restriction, the DNA should solubilize within 30–60 min.

8. The DNA is now ready for another mini-test gel (this one to assess the completeness of the restriction digestion) and then an analytical gel.

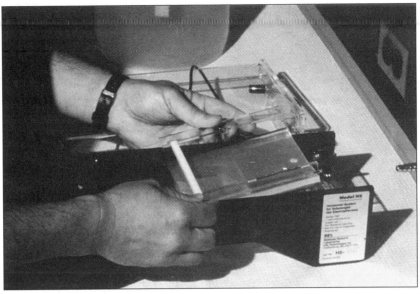

Figure 11. Agarose gel, maintained on plastic template, being placed in electrophoresis apparatus. The tape has been removed. Note that sample well comb is still inserted in the gel.

5.2 TEST GEL TO ASSESS EFFICIENCY OF RESTRICTION ENDONUCLEASE DIGESTION

This gel is known as the post-restriction test gel. It is used to determine the effectiveness of the digestion of DNA by the restriction endonuclease. Documentation is maintained for each post-restriction gel and the DNA samples that have been included on each gel, along with the appropriate controls. The controls include a visual marker consisting of λ DNA *Hind*III-digest and 200 ng of *Hae*III-digested K562 DNA. The visual marker indicates that electrophoresis has proceeded properly. Each aliquot of the digested DNA samples should be visualized as a smear between the largest and smallest fragments of the λ DNA *Hind*III-digest (unless the DNA had been observed on the yield gel to be highly degraded).

The 200 ng *Hae*III-digested K562 DNA is used to compare with the digested DNA. Each sample should resemble the *Hae*III-digested control DNA upon visualization; differences may be an indication that further restriction endonuclease digestion is required or that the extracted DNA was highly degraded. The digested K562 DNA can also be used to determine if further dilution of the specimen is needed. The intensity of the stained sample relative to the 200 ng of digested K562 DNA is used to determine the amount of that sample to be applied to the analytical gel.

The results of the post-restriction gel are photographed.

Materials and Reagents

- Loading buffer solution
- TE buffer

Procedure

1. Vortex-mix the tube briefly, then centrifuge the tube for 5 s at high speed in a microcentrifuge before the top is opened.

2. Remove 2 µL of DNA and combine with 1 µL loading buffer solution in a separate tube. Centrifuge for 2 s.

3. Pipet the entire 3 µL into a test gel well. Run gel under same conditions as described in Chapter 3 entitled Determination of the Quantity of DNA. (See agarose gel setup and protocol for quantitation of agarose gel electrophoresis and ethidium bromide staining.)

4. Completely digested DNA will be present on this test gel as a smooth smear from the dye front back toward the origin. If a fluorescent large band remains near the origin, the digestion is incomplete and should be repeated. To redigest, add 18 µL TE buffer to the 14 µL of remaining DNA to restore the volume to 32 µL.

5.3   SEPARATION OF DNA FRAGMENTS ON AN ANALYTICAL GEL

The analytical gel is designed to separate the *Hae*III-digested DNA. Documentation is maintained for each analytical gel and the DNA samples that have been included on a particular analytical gel. To ensure the integrity of the analytical process, only the samples from one case are applied to an analytical gel. If the number of samples exceeds the capacity of the analytical gel, an additional gel can be created; however, no samples from any other case are included on the additional gel.

The analytical gels contain several controls. They are as follows:

1. Analytical gel visual marker (Adenovirus II DNA-*Kpn*I digest)—consists of nine fragments ranging in size from 1086 to 7713 bp. These fragments are visualized by ethidium bromide staining and are used to determine if the appropriate degree of electrophoresis has occurred.

2. Molecular weight markers that consist of 30 fragments of viral DNA digests ranging in size from 640 to 23 408 bp. These fragments are visualized by hybridization to DNA probes specific for the fragments and autoradiographic or chemiluminescent detection. The molecular weight markers are used to assign a fragment-length value to the control human DNA and specimen DNAs. The appearance of the molecular weight markers on the autoradiographs can be an indicator of proper performance or any anomalies that may have occurred during electrophoresis. These markers should be placed in every four or five sample lanes to minimize electrophoretic distortions that may affect estimation of the size of DNA fragments in unknown samples.

3. Human DNA control (*Hae*III-digested K562 DNA). 400 ng of the digested K562 DNA are applied to each analytical gel. This DNA has been characterized extensively for all the genetic loci examined routinely in the forensic laboratory. The appearance of the VNTR profile results of the K562 DNA, and the fragment-lengths derived from that sample, are noted to ensure that the analytical procedure has generated the expected results for the control DNA.

The results of the analytical gel are photographed and included in the case documentation.

Materials and Reagents

- Gel tray
- Submarine electrophoresis tank (H5 Horizontal Gel Apparatus; Life Technologies)
- *Kpn*I-digested Adenovirus II DNA
- Molecular weight markers
- *Hae*III-digested K562 cell line DNA
- SeaKem® ME (FMC BioProducts, Rockland, ME, USA) or Type II (Sigma Chemical, St. Louis, MO, USA) agarose
- Tris acetate EDTA (TAE) buffer
- Ethidium bromide (EtBr)
- Distilled water
- Agarose gel loading buffer solution
- TE buffer
- Polaroid film No. 553 (400 ASA; Polaroid, Cambridge, MA, USA)

Procedure

The analytical gels are composed of 1% agarose in 1× TAE buffer. The gel dimensions are 11 × 14 × 0.65 cm (100 mL volume).

1. Prepare 1 liter of 1× TAE buffer. Add 100 µL EtBr (10 mg/mL) to the buffer.

2. Prepare the analytical gel:

   a. Weigh out 1.0 g agarose (Sigma Type II or SeaKem ME) into a flask or bottle.

   b. Add 100 mL 1× TAE/EtBr (make mark on flask to indicate volume level.

   c. Bring to a boil to dissolve agarose.

   d. Place at 56°C to equilibrate (if the level of the dissolved agarose solution is below the mark, add distilled water to bring the volume up to its original level).

   e. Place the gel tray on a leveling platform.

   f. Place a 14- or 16-well comb into the gel tray.

   g. Pour agarose into gel form.

h. Let stand at least 15 min to cool.

3. Place the gel into the submarine electrophoresis tank (H5 Horizontal Gel apparatus) with the well comb in place. The buffer should cover the gel to a depth of at least 0.5 cm. Remove the comb.

4. Pour the remaining approximately 900 mL of 1× TAE buffer into the gel tank.

    Note: Well number 1 is defined as the well at the far left side of the gel (and when the wells of the gel are near the analyst). This well will receive visual marker.

5. Incubate analytical gel visual marker (12 μL for each gel) for 5 min at 56°C.

6. Add analytical visual marker to lane 1.

    Note: Well 3 is reserved for the *Hae*III-digested allelic control.

7. Add 400 ng (16 μL) of commercially prepared *Hae*III-digested K562 DNA to lane 3.

    Note: Size markers are placed into the appropriate wells depending upon which gel comb has been used and depending upon the number of samples (reference and unknown) that must be run in the gel.

8. Add 12 μL of molecular weight marker DNA to the appropriate wells.

9. Prepare sample DNAs:

    a. To the 14 μL digested DNA, add 4 μL loading solution.

    b. Mix, centrifuge for 2 s, and carefully pipet the entire specimen into the well.

    c. Repeat for all specimens. If less than 14 μL of the digested DNA is used, add TE buffer to bring the volume to 14 μL.

10. Set the voltage at 30 V (and maximum amperage) for a run time of 17 h. Alternatively, use 32 V (and maximum amperage) for a run time of 16 hours.

11. The analytical gel run is considered complete when the top fragment band of the visual marker has migrated between 9 and 11 cm from the origin.

12. After the electrophoresis is complete, the gel can be examined on the UV transilluminator to evaluate fragment migration/separation. Photograph gels with Polaroid No. 553 (ASA 400) for 1 s at f4.5 with a red filter.

5.4 | SOUTHERN BLOTTING OF DNA ONTO NYLON MEMBRANES

Materials and Reagents

- Oven/incubator

- 0.4 M NaOH
- Soak blot pads (Life Technologies)
- Sponge (Lifecodes, Stamford, CT, USA)
- Biodyne™ B nylon membrane (Pall, Port Washington, NY, USA)
- Glass pipet
- 3 MM chromatography paper (Whatman, Clifton, NJ, USA)
- 0.2 M Tris pH 7.5, 2× standard sodium citrate (SSC)

Procedure

1. Slide the gel from the tray, face down, into a plastic box that contains 0.4 M NaOH sufficient to immerse the gel. Gently shake for 30 min. Ensure that the solution covers the gel and that the gel is not adhering to the bottom of the plastic box.

   a. While gel is incubating in denaturing solution:

   (i) Soak blot pad in a separate container of 0.4 M NaOH for 15 min. Then, discard the NaOH and refill with fresh NaOH for an additional 15 min.

   (ii) Thoroughly soak a sponge with 0.4 M NaOH and place it into a plastic dissecting tray.

   (iii) Cut Biodyne B membrane to the dimensions of 11 × 12.5 cm. Orient the membrane on a surface so that the 11-cm edges represent the top and bottom of the membrane. Label the membrane on the upper left corner.

   Note: Membrane(s) can be cut in advance and stored at room temperature until use.

   b. Fifteen min prior to the end of the gel/DNA denaturation, immerse Biodyne B membrane in 0.4 M NaOH in a separate container for 15 min. The membrane should be handled only with gloved hands.

2. Place the soaked blot pad onto the sponge.

3. Carefully remove the gel from the NaOH solution. Place the gel onto the blot pad, keeping the original gel top face down on the blot pad with the gel origin nearest the analyst. With gloved fingers, press down carefully on the gel to remove any air bubbles.

4. Place the presoaked Biodyne B membrane onto the gel. The label should be located on the bottom right corner (origin end). Ensure that the edges of the membrane are square with the gel edges. Roll a glass pipet up and down the membrane several times to remove any air bubbles.

5. Cover the membrane with a piece of Whatman 3 MM chromatography paper that has been cut to 11 × 12.5 cm and is soaked with 0.4 M NaOH. Roll the surface.

6. Place 9 dry blot pads on top of the Whatman paper.

7. Place two 15 × 20 × 0.4 cm glass plates on top of the dry blot pads.

8. Allow the transfer to proceed for 6 h at room temperature.

9. Remove blot pads and Whatman paper. Grasp the membrane at the right corner (origin end), remove, and turn it over.

10. Wash the membrane once with 0.2 M Tris, pH 7.5, 2× SSC for 15 min with gentle shaking. Blot the membrane on a sheet of Whatman 3 MM chromatography paper.

11. Place each membrane in a folder made from a piece of Whatman 3 MM chromatography paper, 13 × 34 cm, folded lengthwise; tape edges and place in an oven set at 80°C for 30 min.

5.5 RADIOLABELING OF DNA PROBES

Materials and Reagents

- Bench-Count Model BC2000 (NEN Life Science Products, Boston, MA, USA)
- Probe
- Random Priming DNA Labeling System (Life Technologies)
- Water
- dATP, dGTP, and dTTP
- $[^{32}P]$dCTP (50 µCi at 3000 Ci/mmol)
- Klenow fragment
- Spermine-4 HCl
- Herring sperm DNA
- TE buffer
- 5.0 M NaCl

Probes can be labeled with radioisotope using the primer extension method. The quantity of DNA probe taken for labeling varies as a function of the particular probe. Table 2 shows examples of the quantities of probe that can be used.

Procedure

1. Place the appropriate quantity of probe in a 1.5-mL screw-cap tube. Add water to obtain a final volume of 23 µL.

2. Place the tube in boiling water for 8 min. Immediately afterward, place the tube into a slurry of crushed ice and water for 5 min.

3. While on ice, add to the tube:

   a. 2 µL dATP

      2 µL dGTP

      2 µL dTTP

      15 µL random primers buffer mixture

      5.0 µL $[^{32}P]$dCTP (50 µCi at 3000 Ci/mmol)

Table 2. Quantities of Probe That Can Be Used	
Probe (Locus)	**ng used**
YNH24 (D2S44)	100
V1 (D17S79)[a]	100
MS1 (D1S7)[a]	30
PH30 (D4S139)[a]	50
CMM101 (D14S13)	100
TBQ7 (D10S28)	100
LH1 (D5S110)[a]	100
Φ-X174[b]	100
Lambda[b]	100

Probe labeling is carried out using the Random Primers DNA Labeling System.
[a]Purified insert DNA.
[b]Purified inserts from viral DNA.
Unmarked probes still reside in the plasmid vector.

Note: It is very easy to radioactively contaminate the threads of a tube. Once the threads are contaminated, the radioactivity can be transferred to the worker's fingers and hence transferred to anything else that is touched. To avoid such contamination, centrifuge the tube each time its contents are mixed. Check fingers for radioactivity frequently.

   b. Mix.

   c. Add 1 µL Klenow fragment.

   d. Mix and then centrifuge for 2 s.

     50 µL is the final volume.

4. Incubate at room temperature (25°C) for 3 h to overnight.

5. Precipitate the DNA with spermine:

   a. To the 50 µL reaction mixture, add 140 µL TE buffer and 4 µL of 10 mg/mL herring sperm DNA.

   b. Mix.

   c. Add 4 µL 0.1 M spermine-4 HCl.

   d. Place on ice for 15 min.

6. Centrifuge for 10 min in a table top centrifuge at 4°C.

7. Remove supernatant and place in a radioactive waste storage bottle.

8. Add 396 µL TE buffer, followed by 4 µL of 0.1 M spermine to the tube

and vortex-mix briefly.

9. Centrifuge for 2 min at 4°C and remove and discard the supernatant into a radioactive liquid waste storage bottle.

10. To resuspend the pellet, add 520 μL TE buffer and 40 μL 5.0 M NaCl.

11. Mix and place at 56°C for 15–30 min.

12. Vortex-mix the tube briefly. Remove 2 μL labeled probe and place in the exact bottom of a 1.5-mL screw-cap tube. If 2 μL is deposited onto the side of the tube, the disintegrations will not be counted properly. Centrifuge for 2 s if it is necessary to place the 2 μL at the bottom.

13. Place the tube containing 2 μL of probe into the Bench-Count Model BC2000 radioactivity counter. Start the counter and count to 2% precision.

14. Calculate disintegrations per minute (dpm) isotope present in the probe preparation.

   Example: cpm = 10 000

   counting efficiency = 6.8% (instrument specific)
   volume probe counted = 2 μL

   Calculations: $(10\,000\ \text{cpm})(1/0.068)(\frac{1}{2}) = 73529\ \text{dpm/μL}$

   $(73529\ \text{dpm/μL})(560\ \text{μL}) = 4.1 \times 10^7\ \text{dpm total}$
   $(4.1 \times 10^7\ \text{dpm})/(0.1\ \text{μg DNA}) = 4.1 \times 10^8\ \text{dpm/μg}$

15. Calculate probe volume that should be added to a hybridization solution to achieve 5×10^5 dpm/mL solution (Table 3).

The concentration of radioactive viral DNA probe placed into a hybridization solution is a function of the species of DNA. Table 4 illustrates those differences.

5.6 USE OF COMMERCIALLY AVAILABLE RADIOACTIVE DNA PROBES

DNA probes that have been prelabeled with radioisotope can be obtained commercially. This includes probes to the VNTR loci, as well as probes to the viral DNAs that make up the molecular size markers. The requisite dpm for a 60-mL hybridization volume is supplied by the commercial vendor for each of the VNTR locus probes and the DNA marker probes. Documentation regarding the specific activity of each probe supplied and their specific dpm should be provided by the vendor.

Some commercially obtained DNA probes are obtained in combination. That is, one vial will contain both labeled probe for a single VNTR locus as well as probes to the viral molecular weight markers. In other cases, labeled single locus probes will be obtained separately from labeled viral DNA probes. For the latter situation, appropriate volumes of each of the probes are combined into one tube prior to denaturation.

Table 3. Probe Volume Calculations	
Probe (locus)	**dpm/60 mL of hybridization solution**
YNH24 (D2S44)	3.0×10^7
V1 (D17S79)	3.0×10^7
MS1 (D1S7)	3.0×10^7
PH30 (D4S139)	1.5×10^7
TBQ7 (D10S28)	3.0×10^7
LH1 (D5S110)	3.0×10^7
CMM101 (D14S13)	3.0×10^7

5.7 HYBRIDIZATION

Materials and Reagents

- Sterile water
- PEG
- 20× sodium chloride sodium phosphate EDTA (SSPE)
- 20% SDS
- 0.2 M NaOH
- Herring sperm DNA

Procedure

Sixty milliliters of hybridization solution are used in the hybridization containers when hybridizing 11×12.5-cm membranes.

1. Based on the number of containers, prepare the appropriate amount of hybridization solution. For quick reference, use Table 5.

2. Place 60 mL of hybridization solution into each container.

3. Add the membranes to the hybridization solution one at a time, making certain that each is covered with solution before the next is added. As many as 6 membranes can be hybridized in the same container.

4. For each hybridization, set up an individual 15-mL tube that contains 600 µL 0.2 M NaOH and 1.5 mL herring sperm DNA (10 mg/mL).

5. To this tube add the appropriate volumes of single locus probe and viral DNA probes. Mix the tube's contents.

6. Place the tube into boiling water for 5 min.

7. Tilt the container to pool the hybridization solution at one corner and then add the denatured labeled probes (VNTR probe and size marker probes) and rotate the container by hand to mix.

8. Incubate at 65°C overnight with constant shaking.

Table 4. Radioactive Viral DNA Concentrations in Hybridization Solution

Molecular Weight Standard	Viral DNA	dpm/60 mL hybridization solution
LifeCodes (Extended)	ΦX174	1.3×10^6
	Lambda	8.0×10^6

Example: 60 mL hybridization solution
5×10^5 dpm/mL hybridization solution
probe label = 73 529 dpm/μL
(60 mL)(5×10^5 dpm/mL)/(73 529 dpm/μL) = 408 μL

Table 5. Preparation of Appropriate Amounts of Hybridization Solution

No. Containers	Sterile Water (mL)	50% PEG (mL)	20× SSPE (mL)	20% SDS (mL)
1	20.4	12	4.5	21
2	40.8	24	9	42
3	61.2	36	13.5	63
4	81.6	48	18	84
5	102	60	22.5	105
6	122.4	72	27	126
7	142.8	84	31.5	147
8	163.2	96	36	168
9	183.6	108	40.5	189
10	204	120	45	210

For more than 10 containers, add the volumes from above to obtain the volumes. For example, 15 containers would be the volumes for 10 containers plus 5 containers.

5.8 POST HYBRIDIZATION WASHES

Materials and Reagents

- Low-stringency wash
- High-stringency wash (0.1× SSC plus 0.1% SDS)
- 3 MM chromatography paper

Procedure

1. Pour off the hybridization solution slowly. The membranes generally will not slide out of the container during decanting.

2. Perform the following washes, using sufficient wash solution to fill the container halfway.

   a. 15 min in low-stringency wash solution at room temperature.

   b. 15 min in low-stringency wash solution at room temperature.

   c. High-stringency wash solution at 65°C. See Table 6 to determine the appropriate number and length of the high-stringency wash(es) for the various probes.

3. Lightly blot the membrane on Whatman 3 MM chromatography paper.

   Note: **Do not let the membrane dry out!**

5.9 AUTORADIOGRAPHY

Materials and Reagents

- Autoradiography cassette with intensifying screens

- Glad® or Saran® wrap
- Kodak XAR film (Eastman Kodak, Rochester, NY, USA)

Procedure

1. Wrap the damp membranes in Glad or Saran wrap.

   Note: Do not use the Reynolds food wrap for this step because static electricity may be produced. Static electricity can leave images on the film and mask the DNA image.

2. In the darkroom under red light illumination, place the membranes DNA side down (opposite side from membrane identifying label) onto Kodak XAR film. Tape the membranes to this film (place tape only on the edge of the film so as not to mask the patterns). The locations of membranes in contact with the blot are noted by writing directly on the film in one corner with a ball point pen. Place another sheet of XAR film onto the back of the membranes and close the cassette.

Table 6. High-Stringency Washes for Various Probes

Probe (locus)	Number of washes	Length of each wash
YNH24 (D2S44)	1	30 min
V1 (D17S79)	1	30 min
MS1 (D1S7)	2	30 min
PH30 (D4S139)	2	30 min
CMM101 (D14S13)	1	30 min
TBQ7 (D10S28)	2	30 min
LH1 (D5S110)	1	30 min

Note: The high stringency wash (0.1× SSC + 0.1% SDS) used for the final stringency washes must be at 65°C before it is added to the membranes.

Note: The cassette should contain intensifying screens already in place.

3. Place the cassette at -80°C.

4. The XAR film on the back side of the membranes is removed after a 3-day exposure. This back film is to be used as a guide to determine the length of time that the front film needs to be left in place.

5. The front film can be developed at any time after the back film. The exposure time of the front film is determined by the intensity of the DNA profiles and can range from 3 days to 10 days. Exposure times in excess of 10 days generally do not improve the quality of the autoradiograph and should be avoided. If particularly intense DNA profiles are observed, the membrane can be placed on a new piece of film for a shorter exposure period, which could be as short as a few hours.

5.10 BLOT STRIPPING

Materials and Reagents

- Formamide
- 20× SSPE
- 20% SDS
- Distilled water
- Pen

Procedure

1. Remove plastic wrap from membranes.
2. Based on the number of containers, prepare the appropriate amount of blot stripping solution. For quick reference, use Table 7.
3. Add volume of stripping solution to mmbrane.
4. Shake membranes for 90 min at 65°C. Decant the stripping solution.

   Note: To strip the probe for locus D5S110, repeat steps 2 and 3.
5. Place 200 mL of room temperature, high-stringency wash into each box. Rinse for 1–5 min at room temperature.
6. Place the blots onto filter paper to remove excess fluid. Re-label the membrane with a pen if required.
7. Place the blot in the hybridization solution for the next probing.

   Note: If membranes are to be stored for an indefinite period of time, carry out the stripping and rinsing steps as described. Then, rewrap the membranes with plastic wrap and store at room temperature.

Table 7. Preparation of Blot Stripping Solution

No. Boxes	Formamide (mL)	20× SSPE (mL)	20% SDS (mL)	Distilled Water (mL)
1	110	20	10	60
2	220	40	20	120
3	330	60	30	180
4	440	80	40	240
5	550	100	50	300
6	660	120	60	360
7	770	140	70	420
8	880	160	80	480
9	990	180	90	540
10	1100	200	100	600
11	1210	220	110	660
12	1320	240	120	720

For more than 12 containers, add the volumes from the above table to obtain the volumes. For example, 18 containers would be the volumes for 12 containers + 6 containers.

5.11 CHEMILUMINESCENT DETECTION OF RESTRICTION FRAGMENT-LENGTH POLYMORPHISMS

Materials and Reagents

- H5 submarine gel electrophoresis chamber
- Stratalinker® UV Crosslinker (Stratagene, La Jolla, CA, USA)

- Molecular weight marker
- Agarose (Seakem ME or Sigma type II)
- TAE buffer
- 16-well comb
- 0.5 M NaOH, 1.5 M NaCl (denaturation solution)
- Sponges (transfer sponges; LifeCodes)
- Pen
- Blot pad (Life Technologies)
- 1.0 M Tris-HCl, pH 7.5, 1.5 M NaCl (neutralization solution)
- Glass pipet
- 0.2 M Tris, pH 7.5, 2× SSC
- Ziplock® plastic bags

Electrophoresis for Chemiluminescent Detection

Molecular weight markers consist of 30 fragments of viral DNA digests ranging in size from 526 to 22621 bp. These fragments are visualized by hybridization to DNA probes specific for the fragments and chemiluminescent detection. The molecular weight markers are used to assign a fragment-length value to the control human DNA and specimen DNAs. Altered appearance of the molecular weight markers on the lumigraphs can be an indicator of any anomalies that may have occurred during the analytical gel electrophoresis.

Procedure

The analytical gels are composed of 1% low electroendosmosis (EEO) agarose in 1× TAE buffer. The gel dimensions are 11 × 16 cm (100 mL).

1. Prepare 1 L of 1× TAE buffer per analytical gel setup (gel and tank buffer).

2. Prepare the analytical gel:

   a. Weigh out 1.0 g DNA TYPING GRADE™ agarose (Life Technologies) into a flask or bottle.

   b. Place a 16-well comb into the gel tray.

3. Set the voltage at 28 V (maximum amperage) for a run time of 17 h.

4. The analytical gel run is considered complete when the top fragment band of the visual marker has migrated between 10 and 12 cm from the origin. If the top fragment band is not distinctly visible, the 1.699-kb band (second from top) should have migrated between 8 and 10 cm from the origin.

Southern Blotting of Gels onto Nylon Membranes for Chemiluminescent Detection

Procedure

1. Slide the gel from the tray, face down, into a plastic box that contains 0.5 M NaOH, 1.5 M NaCl (denaturation solution) sufficient to cover the gel. Gently shake for 15 min. Ensure that the solution covers the gel, and the gel is not adhering to bottom of plastic box.

   a. While the gel is in denaturation solution:

      (i) Place thin sponges, e.g., transfer sponges, (two per gel) in the transfer trays and add 10× SSC until sponges are saturated and the solution level is just above the bottom sponge.

      (ii) Label membrane with the unique case identifier on the top left corner using a black, U.S. Government Skilcraft™ pen.

   b. Five min prior to end of denaturation soak:

      (i) Place an 11 × 16 cm blot pad in a separate container of 10× SSC.

      Note: Solution will not turn yellow. Blot pad will remain firm.

      (ii) With gloved hands, slowly immerse (to ensure even wetting) an 11 × 14.5 cm Biodyne A membrane (Pall) in 10× SSC in a separate container.

      (iii) Place both containers on an orbital shaker with gentle shaking for the duration of the neutralization step.

2. Rinse gel in deionized or distilled water for 20 s. Gently shake by hand, then decant water. Soak gel in 1.0 M Tris-HCl, pH 7.5, 1.5 M NaCl (neutralization solution) for 15 minutes on an orbital shaker with gentle agitation. Ensure that solution covers the gel, and that the gel is not adhering to the bottom of the plastic box.

3. Place the soaked blot pad on top of sponges. Carefully remove the gel from the neutralization solution. Place the gel onto the blot pad, keeping the original gel top face down on the blot pad with the gel origin nearest to you. With gloved fingers, press down carefully on the gel to remove any air bubbles.

4. Place the presoaked Biodyne A membrane onto the gel, labeled side facing up. Label will be located on the bottom right corner (origin end). Ensure that the edges of the membrane are square with the gel edges. Roll a glass pipet across the membrane to remove any air bubbles from under the membrane.

5. Cover the membrane with a piece of Whatman 3 MM chromatography paper that has been cut to 11 × 14.5 cm and wetted with 10× SSC. Remove air bubbles from between the membrane and the Whatman paper using a glass pipet.

6. Place nine dry blot pads on top of the Whatman paper.

7. Place one $15 \times 20 \times 0.4$ cm glass plate on top of the blot pads.

8. Allow the transfer to proceed until all blot pads are saturated, approximately 4 to 6 h at room temperature. Transfer should not exceed 6 h. Due to the limited volume of 10× SSC, check periodically and add 10× SSC accordingly.

   Note: Do not allow Southern Blot to dry out.

9. Remove blot pads and Whatman paper.

10. Wash the membrane once by placing it in 0.2 M Tris, pH 7.5, 2× SSC, sufficient to cover the membrane for 15 min with gentle agitation. Blot the membrane on a sheet of Whatman 3 MM chromatography paper.

11. Place each membrane in a folder made from a piece of Whatman 3 MM chromatography paper, 13×34 cm, fold lengthwise, tape edges, and place in an oven set at 80°C for 30 min.

12. Open the folder and place the membrane, DNA side facing up (nonlabeled side), in the Stratalinker. Set Stratalinker by pressing Energy, 200 (equivalent to $20\,000$ μJ/cm^2). Press Start. Energy scale will count down to zero. When Stratalinker signals completion, open the door, remove the membrane, and close folder. The membrane can be hybridized at this point or stored in a ziplock plastic bag at room temperature.

Hybridization for Chemiluminescent Detection

Materials and Reagents

- *ACES*™ 2.0 Kit (Life Technologies)
- 1× wash solution
- 0.5× wash solution
- Hybridization solution
- 50-mL plastic conical tube
- Molecular weight marker
- Alkaline phosphatase (AP)-labeled oligonucleotide probe

Procedure

Note: One person should not hybridize more than four boxes at a time, as time and temperature factors are critical.

1. Prepare 500 mL per hybridization box of a 1:10 dilution of *ACES* 2.0 Wash I solution (hereafter referred to as 1× Wash I solution), and 250 mL per hybridization box of a 1:20 dilution of *ACES* 2.0 Wash I solution (hereafter referred to as 0.5× Wash I solution). Preheat solutions to 55°C prior to hybridization. Both solutions can be prepared a day in advance and stored at 55°C until ready for use, or incubate at 55°C approximately 2 h prior to use.

   Note: Preheat wash I concentrate at 55°C prior to making dilution to ensure that all solids are dissolved. If concentrate becomes cloudy, leave at room temperature until clear.

2. Add membranes, DNA side up, to the appropriate volume of *ACES* 2.0 Hybridization solution (see Table 8). Ensure that each membrane is covered with solution prior to adding next membrane. Prehybridize membranes for 20 min in a rotating water bath at 55°C in *ACES* 2.0 Hybridization solution. The rotating speed dial should be set to approximately 60 to 70 rpm. A 50-mL plastic conical bottom tube is used to measure volumes.

Table 8. Volumes for Chemiluminescent RFLP Prehybridizations and Hybridizations	
Volumes for Chemiluminescent RFLP Hybridizations	
No. Membranes	**Prehybridization/Hybridization Volume (mL)**
1-2[a]	30
3-4	
5-6	60
7-8[a]	

[a]Hybridization of 1 to 2 or 7 to 8 membranes may result in reduced band intensity and/or increased membrane background. Four to six membranes should be hybridized at one time.

3. Just prior to the end of prehybridization, microcentrifuge the tubes containing the appropriate chemiluminescent VNTR probe and the molecular weight marker (MWM) probe for 5 min.

   Note: Due to the differences in the relative sensitivities of the AP-tagged oligonucleotide VNTR probes, the recommended hybridization (i.e., probing/stripping) sequence for the VNTR loci is as follows:

   D2S44
   D10S28
   D17S79
   D5S110
   D4S139
   D1S7

   However, a membrane or membranes can be hybridized in any probe sequence, if necessary.

4. Remove the hybridization box from the water bath and decant the prehybridization solution.

5. Add the appropriate volume of VNTR and MWM probe (see Table 9) to a 50-mL conical bottom centrifuge tube containing the appropriate volume of hybridization solution (from Table 8).

6. Vortex-mix briefly and add to the box containing the membranes. Rotate the box gently by hand to distribute the hybridization solution.

   Note: For 60 mL hybridizations, add the probe to 40 mL of solution, vortex-mix gently and add to box with membranes. Add 20 mL of hybridization solution to tube (using tube markings), then add to hybridization box.

7. Hybridize membranes for 20 min in a rotating water bath at 55°C in *ACES* 2.0 Hybridization solution. Ensure that the membranes are not adhering to each other to ensure equal distribution of the probe. The rotating speed dial should be set to approximately 60 to 70 rpm.

   Note: TBQ7 (D10S28) should be hybridized for 30 min under the above conditions.

Post-Hybridization Washes for Chemiluminescent Detection

Materials and Reagents

- Impulse Sealer

- 1× wash solution
- 1× final wash solution
- Lumi-Phos Plus
- Plastic folder

Procedure

1. Decant the hybridization solution from the box. Perform the following washes in the rotating water baths using 250 mL of each wash solution:

   Note: Make sure that membranes are not adhering to each other during all washes to ensure effective washing of all membranes.

   a. 15 min in 1× Wash I solution at 55°C at 60 to 70 rpm.

   b. 15 min in 1× Wash I solution at 55°C at 60 to 70 rpm.

   c. 15 min in 0.5× Wash I solution at 55°C at 60 to 70 rpm.

   d. 5 min in 1× Final Wash solution at room temperature on an orbital shaker with gentle agitation.

   e. 5 min in 1× Final Wash solution at room temperature on an orbital shaker with gentle agitation.

2. Place membranes, DNA side up, on a clean sheet of Whatman 3 MM chromatography paper for 5 to 10 min to draw off excess solution. Do not blot. Place air-dried membranes in a plastic tub containing the appropriate volume of Lumi-Phos Plus (see Table 10). Make sure each mem-

Table 9. Probe Volumes for Chemiluminescent RFLP Hybridizations

	Life Technologies	
Probe for:	Probe volume, 30 mL hybridization[b]	Probe volume, 60 mL hybridization[b]
D1S7[a]	15 μL	30 μL
D4S139[a]	15 μL	30 μL
D5S110[a]	15 μL	30 μL
Molecular weight markers[a]	1.5 μL	3 μL

	Lifecodes	
Probe for:	Probe volume, 30 mL hybridization[b]	Probe volume, 60 mL hybridization[b]
D2S44[a]	15 μL	30 μL
D10S28[a]	30 μL	60 μL
D1S7	30 μL	60 μL
Molecular weight markers	3.8 μL	7.5 μL

	Promega (Madison, WI, USA)	
Probe for:	Probe volume, 30 mL hybridization[b]	Probe volume, 60 mL hybridization[b]
D2S44	30 μL	60 μL
D10S28	30 μL	60 μL
D17S79[a]	120 μL	240 μL

[a]Denotes primary probe source
[b]Volumes may vary depending on quality control results and/or manufacturer's product information.

brane is covered by LUMI-PHOS Plus before adding next membrane. Place tub on rocking platform (set to 3½) for 5 min.

3. Using blunt-end forceps, remove membranes from LUMI-PHOS Plus. Drag membrane alongside of tub to remove excess LUMI-PHOS Plus.

4. Place the membrane in a plastic folder, wipe folder with a Kimwipe to press out air bubbles, and heat-seal folder (impulse sealer set to 5 ½).

5. Trim excess plastic close to outer edge of heat seal and wipe edges with

Table 10. Membrane Number and Appropriate LUMI-PHOS Plus Volumes	
No. Membranes	**LUMI-PHOS Plus Volume**
1–2	15 mL
3–4	20 mL
5–6	25 mL
7–8	30 mL

Kimwipe to remove excess LUMI-PHOS Plus. Two membranes per folder can be accommodated.

Lumography for Chemiluminescent Detection

Materials and Reagents

- X-ray film developer
- Kodak X-Omat RP film (Eastman Kodak)

Procedure

1. The membrane packets are stored overnight at room temperature in the dark, to allow for maximum light output.

2. In the darkroom under red light illumination, place the membrane packets DNA side down onto Kodak X-Omat RP film. Tape the membrane packets to this film. You can record the locations of membranes in contact with the film by writing directly on the film with a ball point pen. Place another sheet of Kodak X-Omat RP onto the back of the membranes and close the cassette. Keep the cassette at room temperature.

3. The Kodak X-Omat RP film on the back side and the front side of the membranes can be developed after an exposure period that is determined for the particular emulsion lot of the Kodak X-Omat RP film in use. This can range from 15 to 60 min for the back film and 30 min to 2 h for the front film. Film exposure times in excess of 2 h are unnecessary, as it will not increase band intensity and will only increase the general background.

Blot Stripping for Chemiluminescent Detection

Materials and Reagents

- Microwave
- 1× strip solution

- 2× SSC
- 3 MM chromatography paper
- Ziplock plastic bags

Procedure

1. Prepare 250 mL 1× strip solution. Heat to 90° to 100°C on stirring hot plate or in microwave.

   Note: Solution will turn cloudy when at appropriate temperature.

2. Cut sealed edges of membrane packet and remove the membrane with blunt-end forceps. Place membranes in a plastic tub containing 250 mL of heated strip solution.

3. Place the tub in a rotating environmental shaker or rotating water bath at 65°C for 15 min. Up to 12 membranes can be stripped in 250 mL.

   Repeat this step once.

4. Pour off strip solution and soak membranes in 250 mL of 2× SSC for 15 minutes at room temperature on an orbital shaker with moderate agitation. Membranes are blotted dry on Whatman 3 MM chromatography paper, then placed in Whatman 3 MM chromatography paper folders and stored in ziplock plastic bags at room temperature to await further hybridizations. Alternatively, the membranes can be left in the plastic folder packet with the LUMI-PHOS Plus, then stripped just prior to subsequent hybridizations.

SUGGESTED READING

1. Adams, D., L. Presley, A. Baumstark, K. Hensley, A. Hill, K. Anoc, P. Campbell, C. McLaughlin et al. 1991. Deoxyribonucleic acid (DNA) analysis by restriction fragment length polymorphisms of blood and other body fluid stains subjected to contamination and environmental insults. J. Forensic Sci. *36*:1284-1298.
2. Amasino, R.M. 1986. Acceleration of nucleic acid hybridization rate by polyethylene glycol. Anal. Biochem. *152*:304-307.
3. Armour, J., S. Povey, S. Jeremiah, and A. Jeffreys. 1990. Systematic cloning of human minisatellites from ordered array charomid libraries. Genomics *8*:510-512.
4. Baird, M., I. Balazs, A. Giusti, G. Miyasaki, L. Nicholas, K. Wexler, E. Kanter, J. Glassberg et al. 1986. Allele frequency distribution of two highly polymorphic DNA sequences in three ethnic groups and its application to the determination of paternity. Am. J. Hum. Genet. *39*:489-501.
5. Balazs, I., M. Baird, M. Clyne, and E. Meade. 1989. Human population genetic studies of five hypervariable DNA loci. Am. J. Hum. Genet. *44*:182-190.
6. Bragg, T., Y. Nakamura, C. Jones, and R. White. 1988. Isolation and mapping of a polymorphic DNA sequence (cTBQ7) on chromosome 10 (D10S28). Nucleic Acids Res. *16*:11395.
7. Budowle, B. and F.S. Baechtel. 1990. Modifications to improve the effectiveness of restriction fragment length polymorphism typing. Appl. Theor. Electrophor. *1*:181-187.
8. Budowle, B., F.S. Baechtel, and D.E. Adams. 1991. Validation with regard to environmental insults of the RFLP procedure for forensic purposes, p. 83-91. *In* M. Farley and J. Harrington (Eds.), Forensic DNA Technology. Lewis Publishers, Boca Raton.
9. Budowle, B., F.S. Baechtel, C. Comey, A. Giusti, and L. Klevan. 1995. Simple protocols for typing forensic biological evidence: chemiluminescent detection for human DNA quantitation and RFLP analyses and manual typing of PCR amplified polymorphisms. Electrophoresis *16*:1559-1567.
10. Budowle, B., J.S. Waye, G.G. Shutler, and F.S. Baechtel. 1990. *Hae*III—a suitable restriction endonuclease for restriction fragment length polymorphism analysis of biological evidence sam-

ples. J. Forensic Sci. *35*:530-536.

11. Church, G.M. and W. Gilbert. 1984. Genomic sequencing. Proc. Natl. Acad. Sci. USA *81*:1991-1995.

12. Denhardt, D.T. 1966. A membrane filter technique for the detection of complementary DNA. Biochem. Biophys. Res. Commun. *23*:641-646.

13. Elder, J. and E. Southern. 1983. Measurement of DNA length by gel electrophoresis. II. Comparison of methods for relating mobility to fragment length. Anal. Biochem. *128*:227-231.

14. Feinberg, A. and B. Vogelstein. 1983. A technique for radiolabeling DNA restriction endonuclease fragments to high specific activity. Anal. Biochem. *132*:6-13.

15. Feinberg, A. and B. Vogelstein. 1983. Addendum. Anal. Biochem. *137*:266-267.

16. Gill, P., A.J. Jeffreys, and D.J. Werrett. 1985. Forensic application of DNA fingerprints. Nature *318*:577-579.

17. Giusti, A., M. Baird, S. Pasquale, I. Balazs, and J. Glassberg. 1986. Application of deoxyribonucleic acid (DNA) polymorphisms to the analysis of DNA recovered from sperm. J. Forensic Sci. *31*:409-417.

18. Giusti, A. and B. Budowle. 1995. A chemiluminescence-based detection system for human DNA quantitation and restriction fragment length polymorphism (RFLP) analysis. Appl. Theor. Electrophor. *5*:89-98.

19. Giusti, A.M. and B. Budowle. 1992. The effect of storage conditions on RFLP analysis of DNA bound to positively charged nylon membranes. J. Forensic Sci. *37*:597-603.

20. Hoopes, B. and W. McClure. 1981. Studies on the selectivity of DNA precipitation by spermine. Nucleic Acids Res. *9*:5493-5504.

21. Jeffreys, A.J., V. Wilson, and S.L. Thein. 1985. Hypervariable minisatellite regions in human DNA. Nature *314*:67-73.

22. Jeffreys, A.J., V. Wilson, and S.L. Thein. 1985. Individual-specific fingerprints of human DNA. Nature *316*:76-79.

23. Johnson, D.A., J.W. Gautsch, J.R. Sportsman, and J.H. Elder. 1984. Improved technique utilizing nonfat dry milk for analysis of proteins and nucleic acids transferred to nitrocellulose. Gene Anal. Tech. *1*:3-8.

24. Kanter, E., M. Baird, R. Shaler, and I. Balazs. 1986. Analysis of restriction fragment length polymorphisms in deoxyribonucleic acid (DNA) recovered from dried bloodstains. J. Forensic Sci. *31*:403-408.

25. Kourilsky, P., O. Mercereau, and G. Tremblay. 1974. Hybridization on filters with competitor DNA in the liquid phase in a standard and micro assay. Biochimie *56*:1215-1221.

26. Laber, T., J. O'Connor, J. Iverson, J. Liberty, and D. Bergman. 1992. Evaluation of four deoxyribonucleic acid (DNA) extraction protocols for DNA yield and variation in restriction fragment length polymorphism (RFLP) sizes under varying gel conditions. J. Forensic Sci. *37*:404-424.

27. Maniatis, T., E.F. Fritsch, and J. Sambrook. 1982. Molecular Cloning: A Laboratory Manual. CSH Laboratory Press, Cold Spring Harbor, NY.

28. McNally, L., R. Shaler, M. Baird, I. Balazs, L. Kobilinsky, and P. DeForest. 1989. Evaluation of deoxyribonucleic acid (DNA) isolated from human bloodstains exposed to ultraviolet light, heat, humidity, and soil contamination. J. Forensic Sci. *34*:1059-1069.

29. McNally, L., R. Shaler, M. Baird, I. Balazs, L. Kobilinsky, and P. DeForest. 1989. The effects of environment and substrata on deoxyribonucleic acid (DNA): the use of casework samples from New York City. J. Forensic Sci. *34*:1070-1077.

30. Milner, E., C. Lotshaw, K. Willems van Dijk, P. Charmley, and H. Schroeder, Jr. 1989. Isolation and mapping of a polymorphic DNA sequence pH30 on chromosome 4 (HGM provisional no. D4S139). Nucleic Acids Res. *17*:4002.

31. Monson, K. and B. Budowle. 1990. A system for semi-automated analysis of DNA autoradiograms, p. 127-132. *In* Proceedings of an International Symposium on the Forensic Aspects of DNA Analyis. U.S. Government Printing Office, Washington, D.C.

32. Nakamura, Y., M. Culver, J. Gill, P. O'Connell, M. Leppert, G. Lathrop, J.-M. Lalouel, and R. White. 1988. Isolation and mapping of a polymorphic DNA sequence pMLJ14 on chromosome 14 (D14S13). Nucleic Acids Res. *16*:381.

33. Nakamura, Y., S. Gillilan, P. O'Connell, M. Leppert, G. Lathrop, J.-M. Lalouel, and R. White. 1987. Isolation and mapping of a polymorphic DNA sequence pYNH24 on chromosome 2 (D2S44). Nucleic Acids Res. *15*:10073.

34. Nakamura, Y., P. O'Connell, M. Leppert, D. Barker, E. Wright, M. Skolnick, M. Lathrop, P. Cartwright, J.-M. Lalouel, and R. White. 1987. A primary genetic map of chromosome 17. Cytogenet. Cell Genet. *46*:668.

35. National Institute of Standards and Technology. 1997. Standard Reference Material 2390 Insert.

36. **Reed, K. and D. Mann.** 1985. Rapid transfer of DNA from agarose gels to nylon membranes. Nucleic Acids Res. *13*:7207-7221.
37. **Sambrook, J., E.F. Fritsch, and T. Maniatis.** 1989. Molecular Cloning: A Laboratory Manual, Vols. 1, 2, and 3. 2nd ed. CSH Laboratory Press, Cold Spring Harbor, NY.
38. **Smith, A.G., C.A. Phillips, E.J. Hahn, and R.J. Leacock.** 1985. Hypersensitization and astronomical tests of X-ray films. AAS Photo-Bulletin *39*:8-14.
39. **Southern, E.M.** 1975. Detection of specific sequences among DNA fragments separated by gel electrophoresis. J. Mol. Biol. *98*:503-517.
40. **Wyman, A.R. and R. White.** 1980. A high polymorphic locus in human DNA. Proc. Natl. Acad. Sci. USA *77*:6754-6758.

PCR-Based Analyses: Allele-Specific Oligonucleotide Assays

6

POLYMERASE CHAIN REACTION (PCR)

An alternative strategy to restriction fragment-length polymorphism (RFLP) typing for forensic DNA typing is the use of PCR-based assays. The PCR has augmented the use of molecular biology for identity testing, as well as for many other fields. The main features that make the PCR especially useful is that it offers greater sensitivity of detection and greater specificity, with the ancillary benefit of speed, than previously used molecular biology procedures.

PCR is an in vitro process that increases the amount of small, specific target sequences; it can be thought of as a form of molecular xeroxing. The salient feature of PCR is the ability to obtain relatively large amounts of specific DNA sequences from relatively small (picogram or nanogram) quantities of genomic DNA. Moreover, analysis takes only a few days for an assay result, and PCR systems are amenable to automation. Generally, successful RFLP typing requires relatively intact DNA (approximately 20 000 bases long). However, many degraded DNA samples can be amplified by PCR and subsequently typed because amplified alleles generally are much smaller in size compared with alleles detected by RFLP/variable number tandem repeat (VNTR) analysis. Thus, PCR is a particularly useful tool for the analysis of forensic material, which may contain degraded DNA due to the age of the sample, environmental exposure, or chemical treatment. In fact, minute quantities of DNA extracted from the following forensic materials are typed routinely and successfully using PCR-based assays in forensic labo-

DNA Typing Protocols: Molecular Biology and Forensic Analysis
By B. Budowle, J. Smith, T. Moretti, J. DiZinno
©2000 Eaton Publishing, Natick, MA

ratories: *(i)* blood, semen, saliva, and sweat deposited on various substrates including clothing, cigarettes, postage stamps, envelope flaps, drinking straws and containers, chewing gum, and face masks; *(ii)* vaginal swabs from a rape victim; *(iii)* various tissues from human remains; and *(iv)* personal items, such as hair brushes, tooth brushes, and razors, which may provide a source of reference samples for identification of unknown remains.

PCR, in principle and often in practice, is a simple process. The essential reagents are comprised of a DNA sample (template), oligonucleotide primers, the four deoxynucleotide triphosphates (dNTPs), buffer, cation, and a thermostable DNA polymerase. All ingredients are mixed in a reaction tube that is inserted into a thermal cycler. The instrument enables a programmable cyclical change in temperature. A typical PCR is based on specific annealing and extension of oligonucleotide primers (two per marker) that flank a defined target DNA segment (Figure 1). Primers are single-stranded DNA oligonucleotides of known sequence, usually 20 to 30 bp in length, that can be obtained commercially or synthesized in-house. The template DNA to be amplified by the PCR is typically denatured by heating the sample to 95°C. After denaturation, at a given locus, each of the two primers hybridize by complementary base pairing to one of the separated strands. Primer annealing is accomplished by lowering the temperature to a defined point (typically between 45° and 65°C). The choice of annealing temperature generally is affected by the sequence of the primers, but is empirically established. The next phase in the PCR cycle, primer extension, is generally carried out at 72°C, the temperature at which *Thermus aquaticus* (*Taq*) DNA polymerase—a commonly used thermostable DNA polymerase—can most effectively copy the original template DNA by extending the primers and making complementary copies of the original template DNA. These three steps (denaturation, primer annealing, and primer extension) represent a single PCR cycle (Figure 1).

Upon repeated cycles of denaturation, primer annealing, and primer extension, an exponential accumulation of a discrete DNA fragment is achieved (Figure 2).

An exponential accumulation of the target DNA sequence occurs because synthesized DNA fragments from previous cycles also serve as templates for amplification. By repeating the cycle, typically 25 to 36 times, at least millions of copies of target sequence can be obtained. Routinely, amplification by PCR can be carried out in this manner in 1 to 2 hours.

The discovery of thermostable DNA polymerases greatly facilitated the PCR. The use of thermostable DNA polymerases, such as *Taq*, was necessitated because of the high temperature required to denature double-stranded DNA prior to copying. Furthermore, the use of a thermostable enzyme has enabled automation of the PCR with thermal cycling instruments. An ancillary benefit of using thermally stable polymerases is that primer annealing and elongation are carried out at high temperatures, which increases sensitivity and specificity of the PCR. Also, at higher temperatures, secondary confirmation of DNA is reduced. More recent developments with the polymerase include a form of *Taq* DNA polymerase, called AmpliTaq Gold™ DNA Polymerase (PE Biosystems, Foster City, CA, USA), that is enzymatically inactive during PCR setup. *Taq* Gold is activated by exposure to high temperature

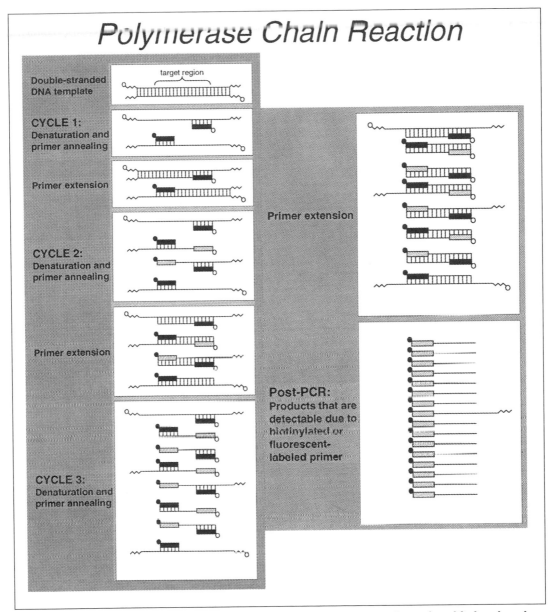

Polymerase Chain Reaction

Double-stranded DNA template

target region

CYCLE 1:
Denaturation and primer annealing

Primer extension

CYCLE 2:
Denaturation and primer annealing

Primer extension

CYCLE 3:
Denaturation and primer annealing

Primer extension

Post-PCR:
Products that are detectable due to biotinylated or fluorescent-labeled primer

Figure 1. Diagram of the PCR process for three cycles. Post-PCR results in an accumulation of amplified product whose size is determined by the position of the primers.

(Figure 3). Thus, the PCR with *Taq* Gold mimics a "hot start" reaction. The result is increased specificity and yield of PCR products (Figure 4).

Because of the nature of forensic samples, contaminants might co-purify with the DNA during extraction. Some of these contaminants may inhibit the PCR. Inhibition of PCR may be overcome by the addition to the reaction of bovine serum albumin (BSA) (Catalog No. A3350; Sigma Chemical, St. Louis, MO, USA) at a concentration of 160 µg/mL. The BSA may bind a

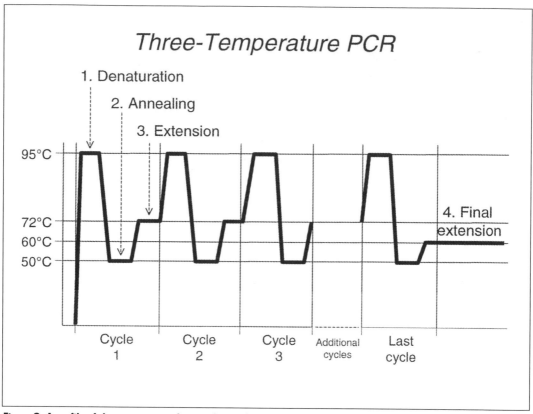

Figure 2. A profile of the temperature changes during the steps of the PCR cycles. Initially the PCR sample is denatured at 95°C, then the temperature is lowered to allow for primer annealing (in this example 50°C), then the temperature is raised to 72°C (a temperature optimum for *Taq* DNA polymerase) for primer extension. After the cycles of PCR are completed, a final extension at 60°C is included to drive the PCR product toward the +A state.

soluble inhibitory factor(s) that co-purifies with DNA, and/or BSA may stabilize the *Taq* DNA polymerase. The source and quality of BSA can impact its effectiveness.

PCR-based analyses involve the amplification of minute quantities of DNA. Thus, precautions are necessary to avoid sample-to-sample contamination. To minimize laboratory-induced DNA contamination, the three main processes of the PCR procedure, which are: (*i*) DNA extraction; (*ii*) PCR setup; and (*iii*) amplified DNA analysis, should be separated by time and/or space. Dedicated equipment, specialized supplies, and special precautions are required for various steps of a PCR-based assay to ensure that reliable results are obtained.

Dedicated pipettor for aliquotting PCR reaction mixture

Dedicated pipettors (adjustable 0.5–10 µL, 2–20 µL, and 20–200 µL) to add components and DNA samples to the PCR mixture

Aerosol resistant sterile pipet tips

Tube racks

Disposable gloves

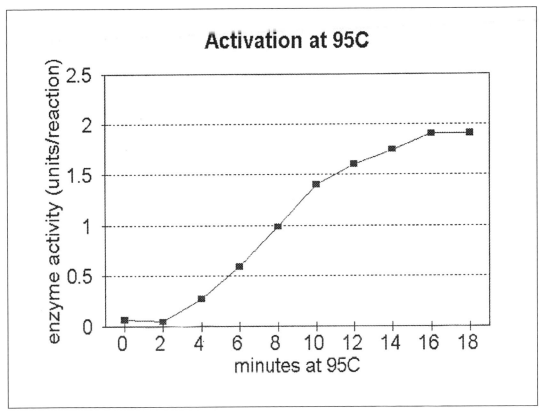

Figure 3. Graph displaying time required for heat activation of AmpliTaq Gold (kindly provided by PE Biosystems, Foster City, CA).

Special precautions for setting up a PCR include:

1. Use disposable gloves and change gloves frequently. Prior to leaving the laboratory area, always remove gloves and wash hands.

2. Use a dedicated 20–200 μL or 2–20 μL pipettor to add sample DNA to the PCR mixture. Never eject the last bit of sample from the pipettor.

3. Always add DNA to the PCR mixture last.

4. After all DNA samples have been added to their respective tubes, close the PCR amplification blank (to which no DNA has been intentionally added) tube last. This tube functions as a negative control for the PCR setup.

5. Avoid touching the inside surface of the tube caps.

6. Change pipet tips after addition of each DNA sample to a PCR mixture.

7. Store the DNA amplification reagents in a refrigerator separate from evidentiary samples.

To avoid the contamination of samples with amplified DNA products (or amplicons), the post-PCR laboratory work area should be a physically separate area from those that involve DNA extraction and PCR setup. The activ-

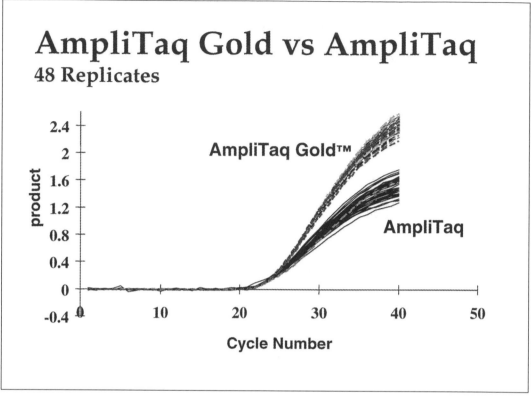

Figure 4. Graph showing increased yield of amplified product using AmpliTaq Gold versus AmpliTaq® (kindly provided by PE Biosystems).

ities in the post-PCR area include: typing of amplified product by hybridization and/or electrophoretic separation, waste disposal of solutions containing amplified DNA, and storage of amplified DNA. Equipment and supplies used to analyze amplified DNA should not be removed from the post-PCR work area.

Dedicated equipment and supplies include:

0.5–10 µL, 2–20 µL, 20–200 µL, and 200–1000 µL adjustable pipettors
Disposable gloves
Microcentrifuge tube de-capping devices
Disposable serological pipets
Towel wipes
Microcentrifuge tube racks
Aerosol resistant pipet tips
Laboratory glassware
Pipet aid

Sink
Source of low vacuum
Aspirator apparatus
Refrigerator
Electrophoresis tanks and power supplies
Orbital shaker platforms
Thermal cyclers
Dispensing pipets
Shaking water baths
Laboratory coats

Pre-PCR (or unamplified) samples should never come into contact with this equipment.

Special precautions for post-PCR sample handling include:

1. Use disposable gloves and change gloves frequently. Prior to leaving the laboratory area, always remove gloves and wash hands.

2. Exercise caution when opening tubes.

3. Use disposable bench paper to cover the work area to prevent the accumulation of amplified DNA on permanent work surfaces. Diluted bleach should be used periodically to wash exposed work surfaces.

4. Use a thermal cycler for amplification or denaturation of amplified DNA. Do not use the thermal cycler for incubation of tubes containing unamplified DNA.

5. The aspirator apparatus is used to collect and isolate waste wash and hybridization solutions generated, and contains amplified DNA. Dispose of these solutions.

6. Store amplified product in the post-PCR area. Do not store amplified samples in refrigerators or freezers in a pre-PCR area.

DNA AMPLIFICATION AND TYPING OF PM AND HLA-DQA1 LOCI

The first post-PCR typing approach used for forensic purposes was detection of sequence polymorphisms by use of allele-specific oligonucleotide (ASO) hybridization probes in a dot blot format. Under appropriate conditions, ASO probes hybridize only to DNA sequences that contain their exact complement. Thus, a different ASO probe is required for each allele to be detected at a locus. A battery of ASO probes is bound to a nylon membrane strip. The configuration where ASO probes are immobilized on a support, instead of amplified DNA, is known as a reverse dot blot format. The strip can accommodate probes for multiple alleles at several loci. The corresponding regions of DNA are amplified by the PCR, and the amplified alleles are hybridized to the immobilized probes to which they are complementary. Because an identifier molecule (or tag) is attached to the 5′ end of one of the primers, a detectable label is incorporated into the amplified alleles. When complexed with probes at fixed locations on the nylon test strip, the amplified alleles can thus be detected and typed (Figures 5 and 6).

Typing of the HLA-DQA1 locus is the most characterized PCR-based system using the reverse dot blot format for the analysis of forensic specimens. The HLA-DQ protein is a heterodimer composed of one alpha chain (encoded by the HLA-DQ alpha locus) and one beta chain. It is expressed in B-lymphocytes, macrophages, thymic epithelium, and activated T-cells. The HLA-DQ protein serves as an integral membrane protein for binding, as well as for presenting, antigen peptide fragments to the T cell receptor of CD4+ T lymphocytes. The polymorphism, which determines the HLA-DQA1 alleles, is detected by amplification and hybridization to the test strip of a 242-bp fragment (or 239-bp length for alleles 2 and 4) from the second exon of the HLA-DQ alpha gene. Eight common alleles have been identified; they are designated 1.1, 1.2, 1.3, 2, 3, 4.1, 4.2, and 4.3. A kit is commercially

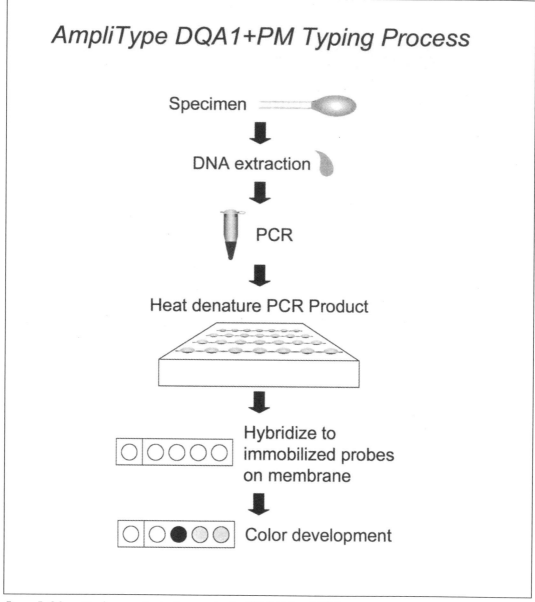

Figure 5. Schematic of process for typing samples for **HLA-DQA1 and PM loci.**

available (AmpliType® PM+DQA1 PCR Amplification and Typing Kit; PE Biosystems) for typing the HLA-DQA1 locus.

Note: Originally a kit was developed to type only the HLA-DQA1 locus, but subsequently more loci can be typed simultaneously using a commercial kit.

Four probes are designed to detect alleles 1, 2, 3, and 4; the 1 allele can be subtyped further as a 1.1, 1.2, or a 1.3 allele, and the 4 allele can be subtyped as a 4.1 or a 4.2/4.3 (the 4.2 and 4.3 alleles cannot be distinguished

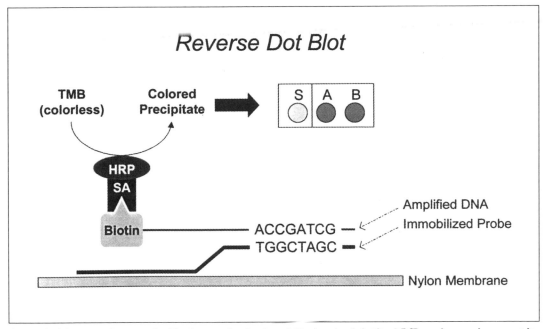

Figure 6. Schematic of reverse dot blot format showing immobilized probe, hybridized PCR product, and streptavidin TMB color detection process.

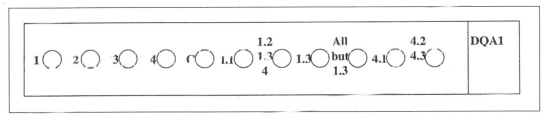

Figure 7. Schematic of array of immobilized probes for HLA-DQA1 reverse dot blot typing.

with the kit). All of the probes for detecting these alleles are contained on a single strip (Figure 7).

The molecular tag attached to one of the HLA-DQA1 primers to detect the amplified allele–probe hybrid complex is biotin. Following hybridization, a streptavidin–horseradish-peroxidase complex is allowed to bind with biotin. The horseradish peroxidase then oxidizes the substrate tetramethylbenzidine (TMB), resulting in a blue precipitate at the hybridization site that indicates the presence of specific alleles.

While the ability to type very small quantities of DNA is possible at the HLA-DQA1 locus, polymorphic data from a single locus does not achieve the power of discrimination provided by RFLP typing of VNTR loci. To increase the discrimination power of PCR-based DNA analyses, the Ampli-Type PM + DQA1 PCR Amplification and Typing Kit also allows for the simultaneous amplification (i.e., multiplex) of the HLA-DQA1 locus and that of five other genetic markers—LDLR, GYPA, HBGG, D7S8, and Gc (Table 1).

Table 1. Characteristics of the HLA-DQA1 and PolyMarker Loci					
Locus	Name	Chromosome Location	No. of Typeable Alleles	Size of Amplicon (bp)	K562 DNA Type
HLA-DQA1	HLA-DQA1	6p21.3	7	242/239	1.1,4.1
LDLR	low density lipoprotein receptor	19p13.1-13.3	2	214	BB
GYPA	glycophorin A	4q28-31	2	190	AB
HBGG	hemoglobin G gammaglobin	11p15.5	3	172	AA
D7S8	anonymous	7q22-31.1	2	151	AB
Gc	group-specific component	4q11-13	3	138	BB

The LDLR, GYPA, HBGG, D7S8, and Gc loci [PolyMarker (PE Biosystems) or PM loci] are typed simultaneously, also using ASO probes by reverse dot blot analysis, in a manner similar to that of HLA-DQA1. With this kit, LDLR, GYPA, and D7S8 each have two detectable alleles (designated A and B), while HBGG and Gc each have three alleles that can be typed (designated A, B, and C) (Figure 8).

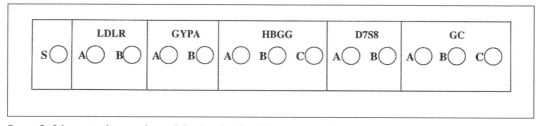

Figure 8. Schematic of array of immobilized probes for PM reverse dot blot typing.

In addition to retaining the useful qualities of the PCR, a multiplex system, such as the DQA1 + PM system, exhibits the following advantages: *(i)* less template (or evidence) DNA is consumed than when analyzing each locus independently; *(ii)* more information is derived than from a single locus analysis; *(iii)* labor is reduced; and *(iv)* because of fewer manipulations, there is less chance of sample contamination or sample mix-up in the laboratory.

The general protocol described above for typing these PCR-based loci entails: extraction of DNA, amplification of specific loci with biotin-labeled primers, denaturation of the amplicons, hybridization of the denatured DNA to probes immobilized on a nylon strip, binding of streptavidin–horseradish peroxidase substrate to the biotin molecules, and detection of allelic products using a colorimetric substrate. While the protocol is straightforward and well-defined, an appreciation for the molecular design of the system can ensure high quality results. For example, the protocol requires transfer of PCR product to the typing trays immediately (i.e., within 30 seconds) following prehybridization denaturation. Extended ambient temperature incubations of denatured Gc B and HLA-DQA1 4.1 products result in a decrease in the B and 4.1 allele signals at the Gc locus and HLA-DQA1 locus, respectively. The probe sites for the Gc B and HLA-DQA1 4.1 alleles are in close proximity to the 3′ end of one of the amplification primers for each locus. Following prehybridization denaturation, unincorporated primers can anneal to their complement if the temperature in the tube reduces to favor annealing conditions. *Taq* DNA polymerase retains activity following the denaturation step and can extend the primer even at ambient temperature. The resulting duplex can block hybridization at the B and 4.1 allele probe sites. The Gc A and C alleles (and other PM alleles) and other HLA-DQA1 alleles are not affected by this primer extension effect because their probe sites are distal to the allele sites and typically duplex formation does not extend to these sites.

Placing samples in the hybridization solution within 15 seconds after denaturation is an effective method to avoid blocking of the HLA-DQA1 and Gc B alleles. If there is any question that this 15-second time frame has been compromised, the sample can simply be denatured again and placed in the hybridization solution within 15 seconds and no loss of Gc B or HLA-DQA1 4.1 allele signals will occur. The addition of EDTA, which chelates the *Taq* DNA polymerase co-factor Mg^{++}, has been proposed as an alternate mechanism to prevent primer extension. Snap-cooling, or immediately placing denatured samples into ice or an ice bath, effectively inhibits primer annealing and substantially reduces *Taq* activity.

6.1   DNA AMPLIFICATION OF PM AND HLA-DQA1 LOCI

Materials and Reagents

- DNA Thermal Cycler (Model 480; PE Biosystems)
- AmpliType PM + DQA1 PCR DNA Amplification and Typing Kit (PE Biosystems; Catalog No. N808-0094)
- PCR tubes
- PCR mixture (in kit)
- Primer mixture (in kit)
- BSA
- Mineral oil
- Deionized water

Procedure

Amplification and typing of the DQA1, LDLR, GYPA, HBGG, D7S8, and GC genes are done using the AmpliType PM + DQA1 PCR DNA Amplification and Typing Kit.

1. Turn on the Model 480 Thermal Cycler (or other model). Using Step

Cycle File No. 4, change the default settings to:

> Denature at 94°C for 1 min.
> Anneal at 60°C for 30 s.
> Extend at 72°C for 30 s.
>
> Program for 32 cycles.
>
> Link to Time Delay File No. 2 for an additional 7-min incubation at 72°C. Link to Soak File No. 1 set at 15°C. Save this file as a User File for later use.

2. Determine the number of samples to be amplified, including controls.

3. If samples, which have been stored at 4°C or frozen, are to be amplified, vortex-mix sample tubes for 10 s and centrifuge in a microcentrifuge for 3 min at maximum speed.

4. Transfer the DNA amplification reagents in the box to the designated PCR setup area. Place PCR mixture tubes that contain 40-µL aliquots of the PM reaction mixture in a rack not used for DNA preparation or amplified DNA handling.

5. Ensure that the solution is at the bottom of each tube by centrifuging very briefly in a microcentrifuge. Label the PCR mixture tubes. Carefully open the tubes. Avoid touching the inside surface of the tube caps.

6. Pipet 40 µL of the primer mixture provided in the kit into each tube, including controls, with a sterile pipet tip. Carefully pipet at a slight angle to minimize mixing and to avoid splashing of solution.

7. Add 2 µL BSA (8 mg/mL) to each tube.

8. Carefully add 2 drops of the mineral oil from the dropper bottle provided in the kit to all tubes including the controls. Do not touch the tube.

9. Use a new, sterile pipet tip for addition of each reagent. Each AmpliType PM PCR amplification is performed in a final volume of 102 µL. Add 20 µL of sample DNA (see guidelines for DNA addition below) to each labeled tube by inserting the pipet tip through the mineral oil layer and expelling the sample. After the addition of the DNA, cap each sample and change pipet tips before proceeding to the next tube. Do not vortex-mix or mix.

   a. Guidelines for DNA addition: add 250 pg to 8 ng DNA to each reaction. Dilute the samples with deionized water to ensure that the desired concentration of DNA is added in a 20 µL volume.

   b. Positive controls: add 20 µL of each 100 ng/mL control DNA to the designated PCR mixture tube.

   c. Negative control: add 20 µL sterile, deionized water to the designated PCR mixture tube.

10. Place the PCR mixture tubes into the thermal cycler. Push the tubes down completely into the heat block. Maintain a record of the heat block position of each tube.

11. Start the thermal cycler amplification program set up in step 1. Verify the cycling parameters by monitoring the first cycle. The tubes should be checked after the first cycle and pressed further into the heat block so that they fit tightly. The PCR amplification program is completed in about 2½ hours.

Figure 9. Tubes placed into the wells of a thermal cycler.

12. After the amplification process, remove the samples from the thermal cycler. Samples are now ready for DNA hybridization and color development, or they may be stored at 4°C for at least 7 days or at -20°C for extended periods. Do not store amplified DNA samples in the same box with unused DNA amplification reagents or DNA samples.

Perform the rest of the assay in amplified DNA work areas using reagents and pipets dedicated for use in this area. Do not transport any of these items to the extraction/PCR setup work area. Also, prior to DNA hybridization, open the tubes one at a time and add 5 µL of 200 mM disodium EDTA. Use a new pipet tip for each addition. Carefully insert the pipet tip through the mineral oil layer. Discard the pipet tip and recap the tube before proceeding to the next tube. The EDTA must be added prior to heat denaturation of the samples, but may be added before or after gel electrophoresis.

6.2 AGAROSE GEL ELECTROPHORESIS EVALUATION OF AMPLIFIED DNA

Materials and Reagents

- Heated stirring plate or microwave

- Minigel apparatus
- Well-forming comb
- Power supply
- UV transilluminator
- Polaroid camera (Polaroid, Cambridge, MA, USA)

- NuSieve® agarose (FMC Bioproducts, Rockland, ME, USA)
- 0.5× Tris-borate EDTA (TBE) buffer
- Ethidium bromide (EtBr)
- Agarose gel loading buffer
- *Hae*III-digested pBR322 plasmid DNA
- Polaroid type 553 or type 55 film (Polaroid)

Procedure

Agarose gel electrophoresis may be performed on the PCR amplified samples to detect amplified DNA. The amplification procedure will yield six products with the following sizes: 242/239 bp (DQA1), 214 bp (LDLR), 190 bp (GYPA), 172 bp (HBGG), 151 bp (D7S8), and 138 bp (GC).

CAUTION: Ethidium Bromide is a mutagen. Wear gloves during this procedure.

1. Prepare 4% NuSieve 3:1 agarose/0.5× TBE–0.5 μg/mL EtBr for minigels using the reagents shown in Table 2.

No. of gels	NuSieve agarose	0.5x TBE–0.5 μg/mL EtBr
1	1.0 g	25 mL
2	2.0 g	50 mL
5	5.0 g	125 mL

Table 2. Minigel Reagents

Place agarose in a bottle. Add 0.5× TBE–0.5 μg/mL EtBr to container.

2. Bring agarose to a boil using a heated stirring plate or microwave oven and continue boiling until agarose is just dissolved.

3. Pour agarose into minigel apparatus using approximately 25 mL/gel. Put in 1 or 2 well-forming combs.

4. When gel is completely sct (approximately 20 min) pour 0.5× TBE–0.5 μg/mL EtBr into gel tank until buffer covers the surface of gel to a depth of approximately 1 mm.

5. While the gel is setting, prepare the samples for electrophoresis as follows. Label enough tubes (0.5 mL or 1.5 mL) for each sample to be evaluated. Place 2 μL loading buffer into each tube. Place 4 μL amplified DNA sample into each tube.

6. Load 2 μL of marker (pBR322 digested with *Hae*III) into the first and last well at each origin. Pipet samples into other wells. Take care to not introduce bubbles or to blow out the sample when loading. Record position of each sample.

7. Connect the electrode leads so that DNA migrates toward the positive electrode. Set the voltage at 200 V. Run until the bromphenol blue has migrated approximately 2–3 cm.

8. Remove the gel from the tank and place on an UV light transilluminator. Photograph the gel using a Polaroid camera and a red filter. Use Polaroid type 553 or type 55 film, f5.6, 1 s exposure.

9. The co-amplified DNA product will reveal the presence of 6 bands that reside between the 104 and 267 bp bands of the size marker. A smaller band and smears, representing primer dimers and unused primers,

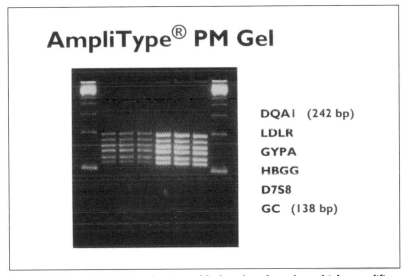

Figure 10. Test gel displaying the six amplified products from the multiplex amplification of the HLA-DQA1 and PolyMarker loci (kindly provided by PE Biosystems).

respectively, may be seen below the 104 bp band of the size marker. The absence of 6 bands between the 104 and 267 bp bands of the size marker indicates a lack of detectable amplified product for all six genetic systems. If samples fail to completely amplify, the PCR may be repeated.

6.3 DNA HYBRIDIZATION

Materials and Reagents

- Rotating water bath
- DNA thermal cycler

- Hybridization solution
- Probe strips
- Typing tray
- Conjugate solution
- Wash solution

The AmpliType PM + DQA1 DNA hybridization process involves 3 steps performed sequentially as follows: *(i)* hybridization of amplified DNA to DNA probe strips, *(ii)* binding of horseradish peroxidase–streptavidin complex to hybridized PCR products, and *(iii)* stringent wash to remove nonspecifically bound PCR products. The color development follows the stringent wash step.

Before starting the DNA hybridization and color development procedures,

Enzyme conjugate (included in kit)	AmpliType PM DNA probe strips (included in kit)
Chromogen: TMB solution (included in kit; prepared in Section 7.4)	Amplitype DQA1 DNA probe strips (included in kit)
Hybridization solution	AmpliType DNA typing trays (PE Biosystems; Part No. N808-0065)
Wash solution	
Citrate buffer	
30% hydrogen peroxide	

assemble the required reagents and equipment as follows:

Clean, disposable gloves should be worn throughout the DNA hybridization and color development steps. Gloves should be discarded when leaving the work area.

Procedure

1. Heat a rotating water bath to 55°C. The temperature should not go below 54°C or above 56°C.

2. The water level should be between 0.5 to 1 cm above the shaker platform. A rotating water bath is necessary for the hybridization and wash steps. Maintain the rotation and the temperature of the water bath throughout the procedure.

3. Warm the hybridization solution and the wash solution to 55°C. All solids must be completely dissolved, and all solutions should be well mixed before use.

4. Using filter forceps, remove the required number of probe strips from the tube. Label each strip in the space at the right edge of the strip. Place one DNA probe strip in each clean well of the typing tray. Strips should all be in the same orientation.

   Note: AmpliType PM and AmpliType HLA DQA1 DNA probe strips can be used to type PCR products from the same PM amplification reaction in parallel, but the DNA probe strips must be placed in separate wells of the tray.

5. Prepare the thermal cycler to denature the amplified DNA by setting the temperature parameter to 95°C. Start the program.

6. Place the tubes in the thermal cycler after it reaches 95°C. Press the

tubes down tightly in the block. Denature the amplified DNA by incubation at 95°C for 3 to 10 min. Keep each tube at 95°C until use.

7. Tilt the typing tray toward the labeled end of the strips. Add 3 mL of prewarmed hybridization solution to each well at the labeled end of each strip. Do not wet the remainder of the strip.

8. Perform the following steps for each tube of amplified DNA within 20 s.

9. Remove a tube from the 95°C block of the thermal cycler.

10. Carefully open the tube and withdraw 20 µL of amplified DNA from the aqueous (bottom) layer and immediately add it below the surface of the hybridization solution into the well of the corresponding probe strip.

11. Cap the tube and set it aside.

12. Repeat steps 9 through 11 until each amplified DNA sample has been added to the corresponding well. Use a new pipet tip for each addition.

    Note: When both AmpliType PM and AmpliType HLA DQA1 DNA Probe Strips are being used to type PCR product from the same tube, add 20 µL of denatured amplified DNA to each of the two strips designated for this sample. The capped tubes can be set aside after the amplified DNA is added to the second DNA probe strip.

13. Put the clear plastic lid on the tray and mix by carefully rocking the tray. Ensure that each strip is completely wet. Once hybridization has begun, strips should remain wet throughout the color development and photography steps.

14. Transfer the tray to the 55°C water bath (stop the rotation of the rotating water bath before adding the tray) and check the temperature. Place a weight on the covered tray to prevent the tray from sliding or floating. Resume rotation of water bath at 50 to 90 rpm.

15. Replace the water bath cover to maintain the temperature at 55°C. Hybridize the amplified samples to the DNA probe strips by incubating at 55°C for 15 min (+ 2 min).

16. Approximately 5 min before the end of the hybridization step, prepare the conjugate solution in a glass flask using the following equations to determine the volume of each component required:

    No. of strips × 3.3 mL prewarmed hybridization solution

    No. of strips × 27 µL enzyme conjugate

    Mix the solution thoroughly and ensure that the solids remain in solution. Leave at room temperature (15° to 30°C) until ready to use.

17. After hybridization, stop the rotation of the water bath and remove the tray. Replace the water bath cover to maintain the temperature at 55°C. Keep the water bath rotating between incubation steps.

18. Aspirate the contents of each well from the labeled end of the strip while tilting the tray slightly. Remove condensation from the tray lid.

19. Dispense 5 mL of prewarmed wash solution into each well. Rinse by rocking the tray for several s, then aspirate the solution from each well.

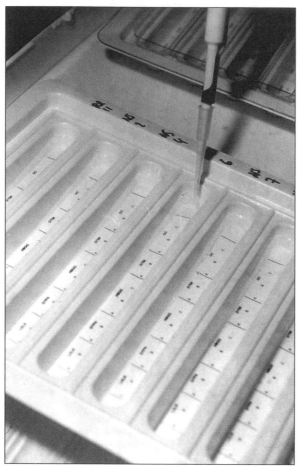

Figure 11. Typing tray containing individual PolyMarker typing strips. Denatured amplified product is being applied to each well.

20. Dispense 3 mL of the enzyme conjugate solution into each well and cover with the clear plastic lid. Stop the rotating water bath and transfer the tray to the 55°C water bath. Place a weight on the covered tray. Adjust the rotating water bath to 50 to 90 rpm.

21. Replace the water bath cover to maintain the temperature at 55°C. Incubate the enzyme conjugate solution with the DNA probe strips at 55°C for 5 min (±1 min).

22. After incubation, remove the tray from the water bath, tip tray at a slight angle, and aspirate the contents of each well from the labeled end of the strips. Remove condensation from the tray lid with a clean laboratory wipe.

    Note: It is important to remove any solution from the tray lid with a clean laboratory wipe at this point to prevent excess blue background coloration from forming during the color development procedure.

23. Dispense 5 mL of prewarmed wash solution into each well. Rinse by rocking the tray for several seconds, then aspirate the solution from each well.

24. To perform the stringent wash step, dispense 5 mL of prewarmed wash solution into each well. Cover the tray with the lid and place it in the 55°C water bath. Place a weight on the covered tray. Adjust the rotation to 50 to 90 rpm.

25. Replace the water bath cover to maintain the temperature at 55°C. Incubate the DNA probe strips at 55°C for 12 min. The temperature and timing of the stringent wash step are critical.

26. After incubation, remove the tray from the water bath. Take off the lid and pour off or aspirate the contents of each well from the labeled end of the strips. Remove any liquid from the tray lid with a clean laboratory wipe.

27. Dispense 5 mL of wash solution into each well. Rock the tray gently for several s.

28. Slowly pour off or aspirate the contents from each well.

6.4 COLOR DEVELOPMENT

Materials and Reagents

- Citrate buffer
- Color development solution
- 30% hydrogen peroxide
- Chromogen solution
- Deionized water

1. Dispense 5 mL of citrate buffer into each well using a dispensing repipet. Cover the tray with the lid, and place it on an orbital shaker set at approximately 50 rpm. Rotate at room temperature (15° to 30°C) for 5 min.

2. During this step, prepare the color development solution. Prepare the color development solution within 10 min prior to use. Add the following reagents in the order listed to a glass flask and mix thoroughly by swirling. Protect from light. Do not vortex-mix. Use the following equations to determine the volumes of each component required:

   No. of strips × 5 mL citrate buffer

   No. of strips × 0.5 μL 30% hydrogen peroxide

   No. of strips × 0.25 mL chromogen solution

3. Remove the tray from the orbital shaker. Remove the lid and slowly pour off or aspirate the contents from each well into a designated waste container. Add 5 mL of the freshly prepared color development solution to each well.

4. Develop the strips at room temperature (15° to 30°C) by rotating on an orbital shaker set at approximately 50 rpm for 20 to 30 min.

Note: Place the lid on the tray and cover with aluminum foil during Steps 4, 6, and 7 to protect the DNA probe strips from strong light.

5. Remove the tray from the shaker. Remove the lid and slowly pour off the contents from each well.

6. Stop the color development by washing the strips in deionized water. Dispense approximately 5 mL of deionized water into each well. Place the tray on an orbital shaker set at approximately 50 rpm for 5 to 10 min. Repeat Step 5.

7. Repeat Step 6 at least two times for a minimum of three deionized water washes. Additional 5–10 min washes will reduce the potential for background color.

6.5 PHOTOGRAPHY AND STORAGE OF STRIPS

Materials and Reagents

- Polaroid camera
- Polaroid type 665 film

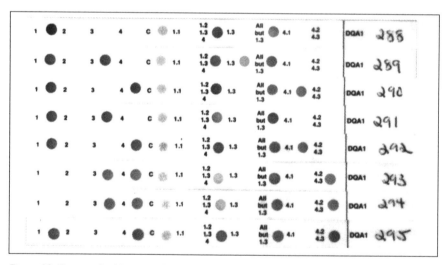

Figure 12. Reverse dot blot strips displaying HLA-DQA1 types. From top to bottom the types are: 1.2,1.2; 1.3,3; 1.2,4.1; 1.2,3; 1.2,4.1; 3,4.2/4.3; 3,4.1; 1.2,4.1.

Procedure

1. Wrap wet strips in plastic wrap. Minimize exposure to strong light.

2. Photograph using Polaroid camera with type 665 film and an orange filter with settings at f8 for 1/8 s.

3. Follow Polaroid film development instructions.

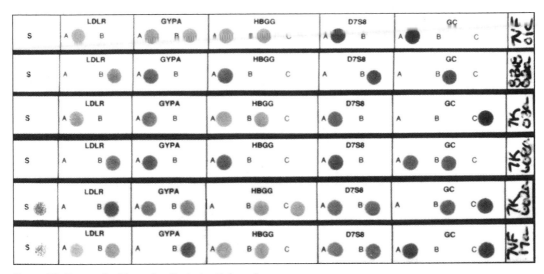

Figure 13. Reverse dot blot strips displaying Polymarker types.

4. Following photography, wrap strips in foil. Protect from light and oxidizing agents (e.g., bleach and nitric acid).

5. Photographs of strip typing results are typically maintained in the case file.

SUGGESTED READING

1. **Allen, M., T. Saldeen, U. Pettersson, and U. Gyllensten.** 1993. Genetic typing of HLA class II genes in Swedish populations: application to forensic analysis. J. Forensic Sci. *38*:554-570.

2. **Blake, E., J. Mihalovich, R. Higuchi, S. Walsh, and H. Erlich.** 1992. Polymerase chain reaction (PCR) amplification and human leukocyte antigen (HLA)-DQα oligonucleotide typing on biological evidence samples: casework experience. J. Forensic Sci. *37*:700-726.

3. **Budowle, B., J.A. Lindsey, J.A. DeCou, B.W. Koons, A.M. Giusti, and C.T. Comey.** 1995. Validation and population studies of the loci LDLR, GYPA, HBGG, D7S8, and Gc (PM loci), and HLA-DQα using a multiplex amplification and typing procedure. J. Forensic Sci. *40*:45-54.

4. **Bugawan, T.L., R.K. Saiki, C.H. Levenson, R.M. Watson, and H.A. Erlich.** 1988. The use of non-radioactive oligonucleotide probes to analyze enzymatically amplified DNA for prenatal diagnosis. Biotechnology *6*:947-953.

5. **Casarino L., F. De Stefano, A. Mannucci, and M. Canale.** 1995. HLA-DQA1 and amelogenin coamplification: A handy tool for identification. J. Forensic Sci. *40*:456-458.

6. **Cimino, G.D., K.C. Metchette, J.W. Tessman, J.E. Hearst, and S.T. Isaacs.** 1991. Post-PCR sterilization: a method to control carryover contamination for the polymerase chain reaction. Nucleic Acids Res. *19*:99-107.

7. **Comey, C.T. and B. Budowle.** 1991. Validation studies on the analysis of the HLA-DQ alpha locus using the polymerase chain reaction. J. Forensic Sci. *36*:1633-1648.

8. **Comey, C.T., B. Budowle, D.E. Adams, A.L. Baumstark, J.A. Lindsey, and L.A. Presley.** 1993. PCR amplification and typing of the HLA-DQα gene in forensic samples. J. Forensic Sci. *38*:239-249.

9. **Comey, C.T., J.M. Jung, and B. Budowle.** 1991. Use of formamide to improve PCR amplification of HLA-DQα sequences. BioTechniques *10*:60-61.

10. **Crouse, C.A., W.J. Feuer, D.C. Nippes, S.C. Hutto, K.S. Barnes, D. Coffman, S.H. Livingston, L. Ginsberg, and D.E. Glidewell.** 1994. Analysis of HLA DQ alpha gene and genotype frequencies in populations from Florida. J. Forensic Sci. *39*:731-742.

11. **Crouse, C.A., V. Vincek, and B.K. Caraballo.** 1994. Analysis and interpretation of the HLA DQA1 1.1 weak-signal observed during the PCR-based typing method. J. Forensic Sci. *39*:41-51.

12. DeStefano, F., L. Casarino, A. Mannucci, L. Delfino, M. Canale, and G.B. Ferrara. 1992. HLA DQA1 allele and genotype frequencies in a Northern Italian population. Forensic Sci. Int. *55*:59-66.

13. Dieffenbach, C.W. and G.S. Dveksler. 1993. Setting up a PCR laboratory. PCR Methods Appl. (Suppl.) *3*:S2-S7.

14. Dragon, E. 1993. Handling reagents in the PCR laboratory. PCR Methods Appl. (Suppl.) *3*:S8-S9.

15. Eckert, K.A. and T.A. Kunkel. 1990. High fidelity DNA synthesis by the *Thermus aquaticus* DNA polymerase. Nucleic Acids Res. *18*:3739-3744.

16. Eckert, K.A. and T.A. Kunkel. 1991. DNA polymerase fidelity and the polymerase chain reaction. PCR Methods Appl. *1*:17-24.

17. Ellingboe, J. and U. Gyllensten (Eds.). 1992. The PCR technique. *In* DNA Sequencing, Eaton Publishing, Natick, MA.

18. Erlich, H.A. (Ed.). 1992. PCR Technology: Principles and Applications for DNA Amplification. Freeman and Company, NY.

19. Erlich, H.A. and T.L. Bugawan. 1992. HLA Class II gene polymorphism: DNA typing, evolution, and relationship to disease susceptibility, p. 193-208. *In* PCR Technology, Principles and Applications for DNA Amplifications. Freeman and Company, NY.

20. Erlich, H.A., D. Gelfand, and J.J. Sninsky. 1991. Recent advances in the polymerase chain reaction. Science *252*:1643-1651.

21. Erlich, H.A., R. Higuchi, K. Lichtenwalter, R. Reynolds, and G. Sensabaugh. 1990. Reliability of the HLA-DQα PCR-based oligonucleotide typing system. J. Forensic Sci. *35*:1017-1018.

22. Erlich, H.A., E.L. Sheldon, and G. Horn. 1986. HLA typing using DNA probes. Biotechnology *4*:975-981.

23. Fildes, N. and R. Reynolds. 1995. Consistency and reproducibility of AmpliType PM results between seven laboratories: field trial results. J. Forensic Sci. *40*:279-286.

24. Gasparini, P., A. Savoia, P.F. Pignatti, B. Dallapiccola, and G. Novelli. 1989. Amplification of DNA from epithelial cells in urine. N. Engl. J. Med. *320*:809.

25. Gyllensten, U. and M. Allen. 1991. PCR-based HLA class II typing. PCR Methods Appl. *1*:91-98.

26. Gyllensten, U.B. and H.A. Erlich. 1988. Generation of single-stranded DNA by the polymerase chain reaction and its application to direct sequencing of the HLA-DQ alpha locus. Proc. Natl. Acad. Sci. USA *85*:7652-7656.

27. Gyllensten, U.B. and H.A. Erlich. 1989. Ancient roots for polymorphism at the HLA-DQα locus in primates. Proc. Natl. Acad. Sci. USA *86*:9986-9990.

28. Haff, L., J.G. Atwood, J. DiCesare, E. Katz, E. Picozza, J.F. Williams, and T. Woudenberg. 1991. A high-performance system for automation of the polymerase chain reaction. BioTechniques *10*:102-112.

29. Harrington, C.S., V. Danuiski, K.E. Williams, and C. Fowler. 1991. HLA DQα typing of forensic specimens by amplification restriction fragment polymorphism (ARFP) analysis. Forensic Sci. Int. *51*:147-157.

30. Hartley, J. and A. Rashtchian. 1993. Dealing with contamination: enzymatic control of carryover contamination in PCR. PCR Methods Appl. (Suppl.) *3*:S10-S14.

31. Hayes, J.M., B. Budowle, and M. Freund. 1995. Arab population data on the PCR-based loci: HLA-DQA1, LDLR, GYPA, HBGG, D7S8, Gc, and D1S80. J. Forensic Sci. *40*:888-892.

32. Helmuth, R., N. Fildes, E. Blake, M.C. Luce, J. Chimera, R. Madej, C. Gorodezky, M. Stoneking et al. 1990. HLA-DQ alpha allele and genotype frequencies in various human populations, determined by using enzymatic amplification and oligonucleotide probes. Am. J. Hum. Genet. *47*:515-523.

33. Herrin, G., N. Fildes, and R. Reynolds. 1994. Evaluation of the AmpliType PM DNA test system on forensic case samples. J. Forensic Sci. *39*:1247-1253.

34. Hochmeister, M.N., B. Budowle, U.V. Borer, and R. Dirnhofer. 1993. Effects of nonoxinol-9 on the ability to obtain DNA profiles from postcoital vaginal swabs. J. Forensic Sci. *38*:442-447.

35. Hochmeister, M.N., B. Budowle, U.V. Borer, and R. Dirnhofer. 1994. Swiss population data on the loci HLA-DQα, LDLR, GYPA, HBGG, D7S8, Gc, and D1S80. Forensic Sci. Int. *67*:175-184.

36. Hochmeister, M.N., B. Budowle, U.V. Borer, and R. Dirnhofer. 1995. A method for the purification and recovery of genomic DNA from an HLA DQA1 amplification product and its subsequent amplification and typing with the Amplitype PM PCR Amplification and Typing Kit. J. Forensic Sci. *40*:649-653.

37. Hochmeister, M.N., B. Budowle, U.V. Borer, U.T. Eggmann, C.T. Comey, and R. Dirnhofer. 1991. Typing of DNA extracted from compact bone tissue from human remains. J. Forensic Sci.

36:1649-1661.

38. Hochmeister, M.N., B. Budowle, U.V. Borer, O. Rudin, M. Bonnert, and R. Dirnhofer. 1995. Confirmation of the identity of human skeletal remains using multiplex PCR amplification and typing kits. J. Forensic Sci. *40*:701-705.

39. Hochmeister, M.N., B. Budowle, J. Jung, U.V. Borer, C.T. Comey, and R. Dirnhofer. 1991. PCR-based typing of DNA extracted from cigarette butts. Int. J. Legal Med. *104*:229-233.

40. Horn, G.T., B. Richards, J.J. Merrill, and K.W. Klinger. 1990. Characterization and rapid diagnostic analysis of DNA polymorphisms closely linked to the cystic fibrosis locus. Clin. Chem. *36*:1614-1619.

41. Huang, N.E. and B. Budowle. 1995. Chinese population data on the PCR-based loci HLA-DQα, LDLR, GYPA, HBGG, D7S8, and Gc. Hum. Hered. *45*:34-40.

42. Impraim, C.C., R.K. Saiki, H.A. Erlich, and R.L. Teplitz. 1987. Analysis of DNA extracted from formalin-fixed, paraffin-embedded tissues by enzymatic amplification and hybridization with sequence-specific oligonucleotides. Biochem. Biophys. Res. Commun. *142*:710-716.

43. Jung, J.M., C.T. Comey, D.B. Baer, and B. Budowle. 1991. Extraction strategy for obtaining DNA from bloodstains for PCR amplification and typing of the HLA-DQα gene. Int. J. Legal Med. *104*:145-148.

44. Kirby, L.T. (Ed.). 1990. DNA Fingerprinting: An Introduction. Stockton Press, NY.

45. Kloosterman, A.D., B. Budowle, and E.L. Riley. 1993. Population data of the HLA DQα locus in Dutch Caucasians. Comparison with seven other population studies. Int. J. Legal Med. *105*:233-238.

46. Koh, C.-L. and D.G. Benjamin. 1994. HLA-DQα genotype and allele frequencies in Malays, Chinese, and Indians in the Malaysian population. Hum. Hered. *44*:150-155.

47. Krawckak, M., J. Reiss, J. Schmidtke, and U. Rosler. 1989. Polymerase chain reaction: replication errors and reliability of gene diagnosis. Nucleic Acids Res. *17*:2197-2201.

48. Kunkel, T. 1992. DNA replication fidelity. J. Biol. Chem. *267*:18251-18254.

49. Kwok, S. and R. Higuchi. 1989. Avoiding false positives with PCR. Nature *339*:237-238.

50. Lareu, M.V., I. Munoz, M.S. Rodriguez, C. Vide, and A. Carracedo. 1993. The distribution of HLA DQA1 and D1S80 (PMCT118) alleles and genotypes in the population of Galicia and central Portugal. Int. J. Legal Med. *106*:124-128.

51. Lee, H.C., E.M. Pagliaro, K.M. Berka, N.L. Folk, D.T. Anderson, G. Ruano, T.P. Keith, and P. Phipps et al. 1991. Genetic markers in human bone I: deoxyribonucleic acid (DNA) analysis. J. Forensic Sci. *36*:320-330.

52. Longo, M.C., M.S. Berninger, and J.L. Hartley. 1990. Use of uracil DNA glycocylase to control carry-over contamination in polymerase chain reactions. Gene *93*:125-128.

53. Lorente, M., J.A. Lorente, M.R. Wilson, B. Budowle, and E. Villanueva. 1994. Sequential multiplex amplification (SMA) of genetic loci: a method for recovery template DNA for subsequent analyses of additional loci. Int. J. Legal Med. *107*:156-158.

54. PE Biosystems. 1990. AmpliType® HLA DQA1 Forensic DNA Amplification and Typing Kit Package Insert. Foster City, CA.

55. PE Biosystems. 1994. AmpliType® PM PCR Amplification and Typing Kit Users Manual. Foster City, CA.

56. PE Biosystems. 1990. AmpliType® User Guide. Foster City, CA.

57. Potsch, L., U. Meyer, S. Rothschild, P.M. Schneider, and C. Rittner. 1992. Application of DNA techniques for identification using human dental pulp as source of DNA. Int. J. Legal Med. *105*:139-143.

58. Presley, L.A., A.L. Baumstark, and A. Dixon. 1993. The effects of specific latent fingerprints and questioned document examinations on the amplification and typing of the HLA DQα gene region in forensic casework. J. Forensic Sci. *38*:1028-1036.

59. Presley, L.A. and B. Budowle. 1994. The application of polymerase chain reaction (PCR) based technologies to forensic analyses, p. 259-276. *In* H.G. Griffin and A.M. Griffin (Eds.), PCR Technology: Current Innovations. CRC Press, Boca Raton.

60. Presley, L.A., J.A. Lindsey, A. Baumstark, A. Dixon, C.T. Comey, and B. Budowle. 1992. The implementation of polymerase chain reaction (PCR) HLA DQ alpha typing by the FBI Laboratory, p. 245-269. *In* Proceedings from the Third International Symposium on Human Identification 1992. Promega Corporation, Madison, WI.

61. Prinz, M., C. Grellner, K. Skowasch, P. Wiegand, B. Budowle, and B. Brinkmann. 1993. DNA typing of urine samples following several years of storage. Int. J. Legal Med. *106*:75-79.

62. Reynolds, R., G. Sensabaugh, and E. Blake. 1991. Analysis of genetic markers in forensic DNA samples using the polymerase chain reaction. Anal. Chem. *63*:2-15.

63. Roy, R. and R. Reynolds. 1995. AmpliType PM and HLA DQα typing from pap smear, semen smear, and postcoital slides. J. Forensic Sci. *40*:266-269.

64. Saiki, R.K., T.L. Bugawan, G.T. Horn, K.B. Mullis, and H.A. Erlich 1986. Analysis of enzymatically amplified β-globin and HLA-DQα DNA with allele-specific oligonucleotide probes. Nature *324*:163-166.

65. Saiki, R.K., D.H. Gelfand, S. Stoffel, S.J. Scharf, R. Higuchi, G.T. Horn, K.B. Mullis, and H.A. Erlich. 1988. Primer-directed enzymatic amplification of DNA with a thermostable DNA polymerase. Science *239*:487-491.

66. Saiki, R.K., S. Scharf, T. Faloona, K.B. Mullis, G.T. Horn, H.A. Erlich, and N. Arnheim. 1985. Enzymatic amplification of beta-globin genomic sequences and restriction analysis for diagnosis of sickle cell anemia. Science *230*:1350-1354.

67. Saiki, R.K., S. Walsh, C.H. Levenson, and H.A. Erlich. 1989. Genetic analysis of amplified DNA with immobilized sequence-specific oligonucleotide probes. Proc. Natl. Acad. Sci. USA *86*:6230-6234.

68. Sajantila, A., M. Strom, B. Budowle, P.J. Karhunen, and L. Peltonen. 1991. The polymerase chain reaction and post-mortem forensic identity testing: application of amplified D1S80 and HLA-DQ alpha loci to the identification of fire victims. Forensic Sci. Int. *51*:23-34.

69. Salazar, M., J. Williamson, and D.H. Bing. 1994. Genetic typing of the DQA1*4 alleles by restriction enzyme digestion of the PCR product obtained with the DQ alpha Amplitype kit. J. Forensic Sci. *39*:518-525.

70. Schneider, P.M. and C. Rittner. 1993. Experience with the PCR-based HLA-DQA1 DNA typing system in routine forensic casework. Int. J. Legal Med. *105*:295-299.

71. Scholl, S., B. Budowle, K. Radecki, and M. Salvo. 1996. Navajo, Pueblo, and Sioux population data on the loci HLA-DQA1, LDLR, GYPA, HBGG, D7S8, Gc, and D1S80. J. Forensic Sci. *41*:47-51.

72. Siebert, P.D. and M. Fukuda. 1987. Molecular cloning of human glycophorin B cDNA: nucleotide sequence and genomic relationship to glycophorin A. Proc. Natl. Acad. Sci. USA *84*:6735-6739.

73. Slightom, J.L., A.E. Blechl, and O. Smithies. 1980. Human fetal $^G\gamma$- and $^A\gamma$-globin genes: complete nucleotide sequences suggest that DNA can be exchanged between these duplicated genes. Cell *21*:627-638.

74. Sullivan, K.M., P. Gill, D. Lingard, and J.E. Lygo. 1992. Characterization of HLA-DQ alpha for forensic purposes. Allele and genotype frequencies in British Caucasian, Afro-Caribbean and Asian populations. Int. J. Legal Med. *105*:17-20.

75. Tagliabracci, A., R. Giorgetti, A. Agostini, L. Buscemi, M. Cingolani, and S.D. Ferrara. 1992. Frequency of HLA DQA1 alleles in an Italian population. Int. J. Legal Med. *105*:161-164.

76. Tahir, M.A. and S. Watson. 1995. Typing of DNA HLA-DQA1 alleles extracted from human nail material using polymerase chain reaction. J. Forensic Sci. *40*:634-636.

77. Tamaki, K., T. Yamamoto, R. Uchihi, Y. Katsumata, K. Kondo, S. Mizuno, A. Kimura, and T. Sasazuki. 1991. Frequency of HLA-DQA1 alleles in the Japanese population. Hum. Hered. *41*:209-214.

78. Tindall, K.R. and T.A. Kunkel. 1988. Fidelity of DNA synthesis by the *Thermus aquaticus* DNA polymerase. Biochemistry *27*:6008-6013.

79. Uchihi, R., K. Tamaki, T. Kojima, T. Yamamoto, and Y. Katsumata. 1992. Deoxyribonucleic acid (DNA) typing of human leukocyte antigen (HLA)-DQA1 from single hairs in Japanese. J. Forensic Sci. *37*:853-859.

80. Walsh, P.S., H.A. Erlich, and R. Higuchi. 1992. Preferential PCR amplification of alleles: mechanisms and solutions. PCR Methods Appl. *1*:241-250.

81. Walsh, S., N. Fildes, A.S. Louie, and R. Higuchi. 1991. Report of the blind trial of the Cetus AmpliType HLA DQα forensic deoxyribonucleic acid (DNA) amplification and typing kit. J. Forensic Sci. *36*:1551-1556.

82. Woo, K.M. and B. Budowle. 1995. Korean population data on the PCR-based loci LDLR, GYPA, HBGG, D7S8, Gc, HLA-DQA1, and D1S80. J. Forensic Sci. *40*:645-648.

Electrophoretic Separation and Allele Detection by Silver Staining for the Loci D1S80 and Amelogenin

7

The discrimination power of variable number tandem repeats (VNTRs) has been combined with the sensitivity and specificity of the polymerase chain reaction (PCR) in an assay termed amplified fragment length polymorphism (AMP-FLP) typing. The process involves similar steps to other PCR-based assays, such as extraction of DNA and amplification of specific sites with designated primers. However, typing of the alleles at a locus is different compared with allele-specific oligonucleotide (ASO) reverse dot blot hybridization assays (described in Chapter 6, entitled PCR-Based Analyses: Allele-Specific Oligonucleotide Assays). The VNTR polymorphism is based on size due to differences in the number of repeat sequences contained within an allele. Because the PCR exponentially amplifies the specified genomic target, sufficient product is available in most cases such that detection with a probe is not necessary (if primer specificity is high). Therefore, the amplified products can be typed by electrophoretic separation of the alleles and subsequent detection of the alleles by using a general stain.

POLYACRYLAMIDE GEL ELECTROPHORESIS

Currently, the most effective manner to resolve VNTR allelic differences in AMP-FLPs is by electrophoresis. Electrophoresis has been and continues to be a cornerstone in typing of DNA variants. The general approach for the analysis of VNTRs had been by agarose gel electrophoresis with subsequent

DNA Typing Protocols: Molecular Biology and Forensic Analysis
By B. Budowle, J. Smith, T. Moretti, J. DiZinno
©2000 Eaton Publishing, Natick, MA

ethidium bromide staining, or by subsequent hybridization assays with isotopic probes. However, agarose gels do not provide the necessary resolution for typing AMP-FLPs. Also, there are more sensitive nonmutagenic stains, such as silver, available for detecting DNA other than ethidium bromide, but the sensitivity of detection by silver staining can be compromised with agarose gels. Additionally, to make technology transfer easier, it would be desirable to eliminate, or at least reduce, the need for potentially hazardous chemicals such as ethidium bromide and radioisotopes.

The qualities of polyacrylamide gel that make it desirable as an electrophoretic medium for AMP-FLP typing are: (*i*) polyacrylamide is inert and, thus, will generally not interact with the DNA molecules; (*ii*) polyacrylamide is charge-free and, thus, the separated components are not usually subjected to electroendosmotic effects; (*iii*) the pore size of polyacrylamide gel can be readily manipulated by changing the monomer and co-monomer concentrations and, thus, the resolution of molecules can be manipulated by sieving effects; (*iv*) polyacrylamide gel is compatible with a wide variety of buffers, so electrophoretic resolution and run times can be manipulated further; (*v*) polyacrylamide gels can be cast as thin as 100 microns, thus enabling separations at higher field strengths due to efficient heat dissipation; and (*vi*) polyacrylamide gel is a clear medium, which makes it suitable for post-separation staining and densitometric assays.

Polyacrylamide gel is the polymerization product of the monomer acrylamide ($CH_2=CH-CO-NH_2$) and a cross-linking co-monomer, such as N,N′-methylene-bis-acrylamide ($CH_2-CH-CO-NH-CO-CH=CH_2$) (BIS). Polymerization is initiated by free-oxygen radicals produced by a base-catalyzed reaction, usually with ammonium persulfate and tetramethylethylenediamine (TEMED). A three-dimensional network of random polymer coils is formed by the cross-linking (via bisacrylamide) of polyacrylamide chains lying side-by-side through the mechanism of vinyl polymerization.

Typically, vertically cast polyacrylamide gels are used for AMP-FLP typing. These gels can enable long separation distances, and because they are sandwiched between glass plates, are usually not subjected to humidity changes in the laboratory. BIS is traditionally used as a cross-linker in polyacrylamide gels. However, piperazine diacrylamide (PDA) can be substituted on a gram-for-gram basis for BIS. It appears that polyacrylamide gels containing PDA have less background staining from silver than BIS containing gels. Thus, the sensitivity of detection of AMP-FLPs in PDA cross-linked gels is enhanced, and it is then possible to store silver-stained PDA gels longer.

Because the charge-to-mass ratio of DNA molecules is the same, regardless of size, DNA components are resolved by altering the effective pore size of the gel. However, size and resolution limits of the electrophoresis of biomolecules can be improved markedly by exploiting the different electrochemical properties of buffers. The buffer system most commonly used for the separation of DNA by electrophoresis is Tris-borate-EDTA (TBE), and to a lesser extent Tris-acetate-EDTA (TAE) and Tris-phosphate-EDTA (TPE). Usually, when using TBE, it is placed in the buffer chambers and in the gel, thus forming a continuous buffer system. While these buffers have proven useful for separation of DNA fragments in agarose gels for restriction fragment-length polymorphism (RFLP) analysis, they do not enable more sophisticated manipulations of the electrophoretic separation process.

An alternative buffer system is one in which the electrode and gel buffers differ in ionic species (i.e., a discontinuous electrophoretic system). Discontinuous buffer system approaches can be achieved by using a pH difference between the buffers such that an initial zone concentration and sharpening of the sample improves resolution compared with continuous buffer methods. DNA fragments can also be separated effectively by using a discontinuous buffer system at an initial constant pH. The approach involves two phases for zone sharpening. The first is the result of a conductivity shift between the sample and the gel. The sample components can be sharpened as they enter the gel to migrate from a lower ionic to a higher ionic environment. The second phase of zone sharpening is the result of a moving boundary between the two buffer zones. The voltage gradient across this boundary causes the trailing end of each sample component to migrate faster than the front of the sample component, resulting in the sample components stacking on the boundary. The effect is that the sample components are concentrated and sharpened as the boundary passes through them. Thus, the components begin as narrow bands when they unstack from the boundary due to mobility differences and sieving effects.

One of the first discontinuous buffer systems used to separate DNA fragments was sulfate-borate. The sulfate-borate system has been very useful for separating PCR products (such as AMP-FLPs) in horizontal and vertical polyacrylamide gels. Additionally, this buffer system has been shown to improve resolution and separation times for DNA sequencing compared with a continuous TBE buffer.

Discontinuous buffer systems allow a tailoring of size separations without altering the gel pore size. Anions, such as formate, chloride, acetate, phosphate, and citrate can be used as the leading ions in the discontinuous buffer system. Different leading anions can affect the resolution and run times of the electrophoretic system. Moreover, other trailing ions, such as glycine, B-alanine, serine, tricine, and Good buffers, have been shown to influence mobility and resolution of DNA molecules. Good buffers with low pKs are more effective for unstacking and thus resolving small-sized DNA fragments, while higher pK Good buffers enable better separation of larger DNA fragments. Changing the overall pH of the buffer system can also alter unstacking limits and resolution.

In addition, when discontinuous buffer systems with borate as the trailing ion are used, compounds can be incorporated into the gel to affect resolution and DNA mobility without changing gel pore size. These mobility modifiers include glycerol and monosaccharides. The stacking limits may be affected because certain sugars and polyhydroxy compounds form complexes with borate in discontinuous buffer systems that use borate as a trailing ion. In most cases there is a significant retardation of DNA mobility in the presence of these mobility modifiers.

By using polyacrylamide as the separation medium for AMP-FLPs, a number of stains for DNA detection can be used in concert with the inert gel material. The use of silver staining of DNA fragments in polyacrylamide gels is a substantially less hazardous assay than ethidium bromide staining and provides for the possibility of developing a permanent record of the electrophoretic separation. Moreover, the higher degree of detection sensitivity of silver compared with ethidium bromide makes the detection of both bands of a preferentially amplified heterozygous sample more likely. The silver staining method can be very simple and entails only a few steps, which include: oxidation, silver incubation, reduction, and a stop wash for the reaction. An ancillary benefit of this approach is that the procedure takes only 15 to 30 minutes to complete.

D1S80 LOCUS

The D1S80 locus serves as a good model to demonstrate the AMP-FLP typing assay, and also the locus is one of the best defined VNTR loci that can be amplified by the PCR. The D1S80 locus has a repeat sequence that is 16 bp in length. Alleles vary in size as a function of the number of repeat sequences contained within them. More than 27 different alleles have been observed at the D1S80 locus. The region to amplify by PCR consists of the VNTR and the sequences flanking the VNTR as defined by the positions of the oligonucleotide primer pair relative to the VNTR. Allelic ladders, run adjacent to the amplified samples, enable determination of the alleles present in known and questioned DNA specimens by visual comparison. The D1S80 allelic ladder contains alleles ranging in size from 369 bp (i.e., 14 repeats) to 801 bp (i.e., 41 repeats).

AMELOGENIN LOCUS

As stated in Chapter 6, entitled PCR-Based Analyses: Allele-Specific Oligonucleotide Assays, there are advantages of a multiplex system, such as less template DNA is consumed than when analyzing each locus independently, less reagents are consumed, and labor is reduced. The amelogenin gene is a locus that also has been demonstrated to be valid for forensic applications and potentially may be analyzed in a multiplex fashion with the D1S80 locus. Typing the amelogenin gene enables determination of the sex of the contributor of a biological sample. The following protocol enables co-amplification and simultaneous typing of the D1S80 and amelogenin loci.

7.1 | DNA AMPLIFICATION OF D1S80 AND AMELOGENIN LOCI

Materials and Reagents

- GeneAmp® PCR System 9600 DNA Thermal Cycler (PE Biosystems, Foster City, CA, USA
- Cap installation tool (Catalog No. N801-0438; PE Biosystems)
- Deionized water
- 10× PCR buffer
- Bovine serum albumin (BSA)

- dATP
- dCTP
- dGTP
- dTTP
- Primers
- *Taq* DNA polymerase
- Strip caps (Catalog No. N801-0535; PE Biosystems)

Procedure

Multiplex amplification of genomic DNA for the D1S80 and amelogenin loci is carried out by combining DNA template, 10× buffer, the four deoxyribonucleoside triphosphates, D1S80 primers, amelogenin primers, BSA, sterile water, and DNA polymerase in a PCR tube and subjecting this reaction mixture to a series of controlled temperature changes in a GeneAmp PCR system 9600 thermal cycler. The DNA template amount can range from 400 pg (minimum) to 5 ng (maximum).

1. Turn on the thermal cycler. Program the thermal cycler according to the following instructions.

   a. Select option.

      Run-Create-Edit-Util.

      Using the option key, press once to move the cursor to Create.

      Press enter.

   b. Create program.

      Hold-Cycle-Auto-Meth.

      Using the option key, depress once to move the cursor to Cycle.

      Press enter.

   c. 3 Temperature PCR.

      Press enter.

   d. Set pt #1   Ramp 0:00.

      94.0C   Hold 0:30.

      Press enter to move the cursor to 94.0C. Change to 95.0C by depressing 9 5 0 enter.

      Change Hold 0:30 to 0:10 by depressing 1 0 enter.

   e. Set pt #2   Ramp 0:00.

      55.0C   Hold 0:30.

      Press enter to move the cursor to 55.0C.

      Change to 67.0C by depressing 6 7 0 enter.

      Change Hold 0:30 to 0:10 by depressing 1 0 enter.

   f. Set pt #3   Ramp 0:00.

      72.0C   Hold 0:30.

      Press enter to move the cursor to 72.0C.

Change to 70.0C by depressing 7 0 0 enter.

The default on Hold is correct. Press enter.

g. Total cycles = 25.

Pause during run? No.

Depress 2 7 enter.

The default no is correct. Press enter.

h. Cycl # ???

Run-Store-Print-Home.

Note: Once the amplification program has been keyed into the thermal cycler, it should be stored for future use. Refer to the Users Manual for instructions on storing programs.

i. Select tube   Micro.

Reaction vol? 100 µL.

Change to 50 µL by depressing 5 0 enter.

j. Select print mode.

Off-Cycle-Setpoint.

The default Off is correct. Press enter.

2. Determine the number of samples that will be amplified, including controls. The controls consist of the following:

a. Amplification positive controls. These are known samples previously characterized at the D1S80 genetic locus. They include DNA derived from the standard cell line K562, as well as DNA processed from an internal laboratory standard.

b. Amplification negative control and the reagent blank(s) control.

3. If amplifying samples that have been stored at 4°C or frozen, vortex-mix the sample tubes for 5 s and centrifuge in a microcentrifuge for 1 min at maximum speed.

4. Transfer the DNA amplification reagents to the designated PCR setup area. Place amplification reagent tubes in a rack that is not used for DNA preparation or amplified DNA handling.

5. Label the appropriate number of PCR tubes and place them in the 96-position amplification tray.

6. Prepare the amplification master mixture according to Table 1.

7. Add 30 µL of master mixture to each tube.

8. Add a volume of sterile, deionized water to each tube that is equal to the difference between 20 µL and the volume of the template DNA. For example, if 5 µL of the DNA template will be added, then 20 µL - 5 µL = 15 µL water must be added. In the case where 20 µL of the template will be added, no additional water needs to be added.

9. Add the DNA template, using a new pipet tip for each sample. The DNA template volume cannot exceed 20 µL.

Table 1. Volumes for Varying Numbers of Samples[a] (All volumes are in μL)				
	No. of Samples			
Component	**10**	**20**	**50**	**100**
H_2O[b]	125	250	625	1250
10× Buffer	50	100	250	500
BSA[c]	10	20	50	100
dATP	10	20	50	100
dCTP	10	20	50	100
dGTP	10	20	50	100
dTTP	10	20	50	100
D1S80 Primer A	10	20	50	100
D1S80 Primer B	10	20	50	100
Amelogenin Primer Pair	50	100	250	500
Taq[d]	5	10	25	50
TOTAL	**300**	**600**	**1500**	**3000**

[a]Based on a DNA template volume of 20 μL.
[b]Sterile, deionized water.
[c]BSA, 8 mg/mL.
[d]*Taq* DNA Polymerase.

Note: The amplification master mixture components (without *Taq*) may be aliquoted into batches and stored frozen prior to usage. Upon thawing, add the appropriate volume of *Taq* to complete the reaction.

10. Place the caps on each row of tubes. Seal the caps tightly. The cap installation tool should be used with the strip caps.

11. Gently tap the corners of the PCR rack on the edge of the laboratory bench.

12. Place the rack into the Model 9600 thermal cycler, close and tighten the sample cover, and initiate the amplification process by pressing enter. The program will begin as soon as the sample cover has reached 100°C.

13. After the amplification process, remove the samples from the thermal cycler. Samples are now ready for electrophoretic resolution. The samples can be stored at 4°C for up to 7 days, or at -20°C for longer periods. Do not store amplified DNA in the same refrigerator as unamplified DNA samples or with DNA amplification reagents.

7.2 ELECTROPHORETIC SEPARATION OF AMPLIFIED ALLELES

Materials and Reagents

- Delrin comb
- Fingerprint roller
- 0.4 M NaOH
- Deionized water
- Tap water
- 100% ethanol
- Glue Stick (No. U26; FaberCastell)
- GelBond® (FMC BioProducts, Rockland, ME, USA)
- Acrylamide
- Bisacrylamide or PDA
- 0.24 M Tris-formate
- 10% ammonium persulfate
- TEMED
- Plastic wrap

The amplified allelic products of the genetic locus D1S80, as well as amelogenin, differ from one another in size. To determine the alleles represented in the original genomic DNA, the amplified products are subjected to polyacrylamide gel electrophoresis. Following electrophoretic separation, the amplified allelic products can be made visible by staining the gel with silver.

Preparation of Polyacrylamide Gel

Procedure

1. Prior to each use, the glass plates used for gel preparation must be cleaned thoroughly. Wash each plate with 0.4 M NaOH, rinse with tap water, and follow with a thorough rinse with deionized water. Place the plates in an upright position and rinse with 100% ethanol. Wipe dry with a lint-free laboratory wiper towel. Wash the Delrin comb with water and permit it to dry.

2. To assemble the plates for use, place the long plate on a flat surface with the long axis of the plate parallel to the edge of the laboratory bench. Using a Glue Stick, apply a 1-cm wide strip of glue to the left end of the plate. After putting on gloves, add deionized water to the glass plate and place a 19.5 × 37 cm piece of GelBond—hydrophobic side down—on top of the glass plate. *Note*: The protective cover sheet remains on the Gel-Bond. Press the excess water out by rolling the GelBond with a fingerprint ink roller. The GelBond protective sheet is removed now. Blot away with a laboratory wipe any water that might have gotten onto the GelBond.

3. Place 0.4-mm plastic spacers along each side of the GelBond-covered plate. The ends of the spacers should be even with the glued end of the GelBond. In addition, the ends of the spacers with the foam cushions should be opposite the end of the gel that was glued to the GelBond.

Place the short plate on top of the spacers, with the foam cushions pressed tightly against the top of the short plate. Make certain the spacers are even with the edges of the plates. Place 3 binder clips on each side of the gel cassette. Place the gel cassette on a level surface.

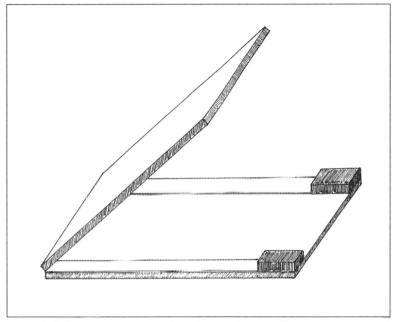

Figure 1. Gel casting requires two glass plates and spacers. The apparatus is placed horizontally for more effective gel casting.

4. Combine the gel ingredients in a beaker.

   7.9 mL acrylamide (30% total acrylamide/2% cross-linker)

   7.0 mL 0.24 M Tris-formate

   13.1 mL sterile, deionized water

   Mix by swirling gently.

   280 μL 10% ammonium persulfate

   28 μL TEMED

   Mix by swirling gently. *Note*: Polymerization begins upon addition of the ammonium persulfate, and TEMED is a catalyst; you must load the gel solution into the cassette immediately.

5. Draw the gel solution into a 25-mL pipet, taking care not to introduce any air bubbles into the pipet, and inject the solution into the cassette from the top of the short plate with the cassette slightly tilted. Continue to load the gel solution until the injection front reaches the bottom of the cassette. The length of time for injection of the gel solution should be no more than 60 s.

6. Immediately insert the comb into the gel and press it tightly against the top edge of the short plate. Permit the gel to polymerize for 1½ h at room temperature. Place the gel at 4°C for at least 30 min prior to use. If gel is to be refrigerated for more than 1 h at 4°C, it should be covered with plastic wrap.

7.3   ELECTROPHORESIS

Materials and Reagents

- Model SA 32 electrophoresis apparatus (Life Technologies, Gaithersburg, MD, USA)
- Syringe
- Power supply
- Loading buffer solution
- Tris-borate buffer
- UniFit disposable, flat microflex tips
- Allelic ladders

Procedure

1. Prepare samples for loading by placing 2.0-µL aliquots of loading solution into a new tube for each sample that will be loaded. To each tube add 4.0 µL amplified DNA sample and mix well.

2. Add 250 mL 28 mM Tris-borate buffer to the bottom tank of the electrophoretic apparatus. Carefully remove the comb from the gel. Install the gel cassette on the electrophoretic apparatus with the long plate to the outside. Tighten the gel clamps firmly and make certain that the drain clamp is closed. Remove any bubbles that may have been trapped at the bottom of the gel. Add 300 mL 28 mM Tris-borate buffer to the top tank of the electrophoretic apparatus.

3. Using a syringe, flush out the wells in the gel by repetitively forcing buffer into the wells.

4. Using UniFit disposable, flat microflex tips, load 5.5 µL sample volume into a well of the gel. The sample volume will be loaded underneath the Tris-borate buffer that will be in the wells. Slowly pipet the sample into the wells so as to form a smooth layer under the buffer.

   Note: The first sample loaded on the gel is the amplification blank. This is followed by the D1S80/amelogenin reference ladder combination, which is loaded on every third lane thereafter across the width of the gel. The positive control samples are loaded into the next two lanes, adjacent to the reference ladders. All samples are then loaded in a manner that allows them to be flanked on either side by a reference ladder.

5. Close the tank covers and attach the electrode wires. Set the power supply to deliver 1000 V. Set the milliamperage and wattage values to their maximums.

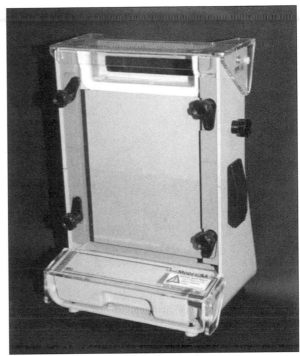

Figure 2. Vertical polyacrylamide gel electrophoresis unit Model SA 32.

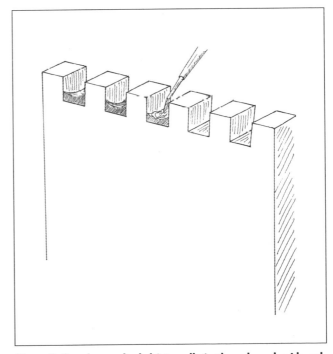

Figure 3. Samples are loaded into wells in the polyacrylamide gel. The wells were formed using a comb during polymerization.

6. Allow the electrophoresis to proceed until the bromophenol blue dye marker line has reached the top of the buffer in the lower tank. Stop the run, drain the buffer from the top tank, and remove the gel from the electrophoresis unit. Pry open the gel cassette using a thin metal spatula and remove the gel from the glass plate. Handle the gel only by the edges of the GelBond. The gel is now ready for staining.

7.4 SILVER STAINING OF GELS

Materials and Reagents

- Plastic tray
- 1% HNO_3
- 0.2% $AGNO_3$
- 0.28 M Na_2CO_3, 0.05% formaldehyde
- 10% glacial acetic acid
- 3% glycerol (vol/vol)
- Filter paper (Whatman, Clifton, NJ, USA)
- GelBond

1. Place the gel (gel side up) into a 32×46 cm plastic dissecting tray, cover the gel with 1% HNO_3, and place the tray on an orbital shaker. Permit

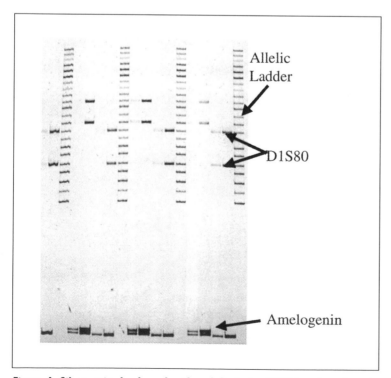

Figure 4. Silver stained polyacrylamide gel displaying D1S80 and amelogenin amplicons.

the gel to shake until the bromophenol blue dye line on the gel (i.e., line that corresponds to the electrophoretic moving boundary) disappears (about 5 min). If necessary, the gel can be left at this stage overnight.

2. Decant the 1% HNO_3 from the tray and rinse the gel and tray copiously with deionized water. Remove the gel and rinse it under a stream of flowing deionized water for 15 to 20 s.

3. Return the gel to the tray and cover it with 500 mL 0.2% $AgNO_3$; place the tray on the orbital shaker and shake for 20 min.

4. Decant the $AgNO_3$ into the dedicated silver recovery vessel and rinse the gel and tray copiously with deionized water.

5. Treat the gel with 0.28 M Na_2CO_3, 0.05% formaldehyde until the DNA bands become suitably visible. Treatment is composed of repetitive rinses and decantations of the development solution. The first several rinses will turn milky brown very quickly and should be decanted immediately.

6. Cover the gel with 10% acetic acid and place on an orbital shaker for 5 min minimum. Note: The gel can be left overnight at this stage if necessary.

7. Rinse the gel in two changes of deionized water. Then cover the gel with deionized water and shake on an orbital shaker for 10 min.

8. Decant the water and cover the gel with 3% glycerol and place on an orbital shaker for 5 min.

9. Remove the gel from the glycerol and place it on a clean piece of Whatman filter paper. With a plastic straightedge, carefully squeegee the excess fluid from the gel surface. Permit the gel to dry at room temperature.

10. After drying, the gel surface can be covered with a clear piece of plastic (e.g., GelBond) and the edges taped shut.

7.5 PHOTODUPLICATION OF SILVER-STAINED GELS

Materials and Reagents

- Light box with even dispersal
- Glass plate
- Film processor
- X-ray duplicating film

The silver-stained gel serves as a permanent record of the separation of the amplicons. However, sometimes the gel integrity may degrade during long term storage—background staining may increase over time, and a gel may crack during storage. Also, while scientists are willing to exchange their results they are, however, reluctant to part with the original dried gels. Standard photography can be used to record the gel. An alternate method for recording the gel is to make a contact print reproduction on X-ray duplicating film of the silver-stained gel. Because X-ray duplicating film contains a photographic emulsion, the duplication process can render a high quality duplicate. Moreover, because of the linear sensitivity range of X-ray duplicating films, good quality results can be obtained from a wide variety of light sources without significant loss of gray scale tonality. Basically, evenly dispersed white light (e.g., 40 W cool white fluorescent bulbs or 15 W incandescent frosted bulbs) is a sufficient light source. The determination of optimal exposure for any light source is derived by a series of empirical test contact prints.

Procedure

1. After analysis, take the dried silver-stained gel (or if still wet, the gel wrapped in plastic wrap) to the dark room.

2. Place the gel on a white light source light box (turned off).

3. Place X-ray duplicating film, emulsion side down, on the light box.

4. Place a glass plate on top of the film to maintain contact between the gel and X-ray film.

5. Turn on light box for a predetermined time (usually 1–10 s) to expose the film.

6. Develop film.

SUGGESTED READING

1. Akane, A., A. Seki, H. Shiono, H. Nakamura, M. Hasegawa, M. Kagawa, K. Matsubara, Y. Nakahori, S. Nagafuchi, and Y. Nakagome. 1992. Sex determination of forensic samples by dual PCR amplification of an X-Y homologous gene. Forensic Sci. Int. *52*:143-148.

2. Akane, A., H. Shiono, K. Matsubara, Y. Nakahori, A. Seki, S. Nagafuchi, M. Yamada, and Y. Nakagome. 1991. Sex identification of forensic specimens by polymerase chain reaction (PCR): two alternative methods. Forensic Sci. Int. *49*:81-88.

3. Allen, R.C. 1971. Polyacrylamide gel electrophoresis with discontinuous voltage gradients, p. 29-32. *In* Electrophoresis and Related Techniques of Polyacrylamide Gel Electrophoresis. Walter de Gruyter, Berlin.

4. Allen, R.C. 1980. Rapid isoelectric focusing and detection of nanogram amounts of proteins from body tissues and fluids. Electrophoresis *1*:32-37.

5. Allen, R.C. and B. Budowle. 1993. The use of the polymerase chain reaction and the detection of amplified products, p. 113-128. *In* B.A. White (Ed.), PCR Protocols: Current Methods and Applications, Methods in Molecular Biology, Vol. 15. Humana Press, Totawa, NJ.

6. Allen, R.C., B. Budowle, and D.J. Reeder. 1993. Resolution of DNA in the presence of mobility modifying polar and non-polar compounds by discontinuous electrophoresis on rehydratable polyacrylamide gels. Appl. Theor. Electrophor. *3*:173-181.

7. Allen, R.C., G. Graves, and B. Budowle. 1989. Polymerase chain reaction amplification products separated on rehydratable polyacrylamide gels and stained with silver. BioTechniques *7*:736-744.

8. Alonso, A., P. Martin, C. Albarran, and M. Sancho. 1993, Amplified fragment length polymorphism analysis of the VNTR locus D1S80 in central Spain. Int. J. Legal Med. *106*:311-314.

9. Alonso, A., P. Martin, C. Albarran, and M. Sancho. 1995. A HinfI polymorphism in the 5' flanking region of the human VNTR locus D1S80. Int. J. Legal Med. *107*:216-218.

10. Baechtel, F.S., J.B. Smerick, K.W. Presley, and B. Budowle. 1993. Multigenerational amplification of a reference ladder for alleles at locus D1S80. J. Forensic Sci. *38*:1176-1182.

11. Bassam, B.J., G. Caetano-Anolles, and P.M. Gresshoff. 1991. Fast and sensitive silver staining of DNA in polyacrylamide gels. Anal. Biochem. *196*:80-83

12. Boerwinkle, E., W. Xiong, E. Fourest, and L. Chan. 1989. Rapid typing of tandemly repeated hypervariable loci by the polymerase chain reaction: application to the apolipoprotein B 3' hypervariable region. Proc. Natl. Acad. Sci. USA *86*:212-216.

13. Budowle, B. and R.C. Allen. 1991. Discontinuous polyacrylamide gel electrophoresis of DNA fragments, p. 123-132. *In* C. Matthew (Ed.), Methods in Molecular Biology, Protocols in Human Molecular Genetics, Vol. 9. Humana Press, Clifton, NJ.

14. Budowle, B. and R.C. Allen. 1998. Protocols for analyzing amplified fragment length polymorphisms (VNTR/STR loci) for human identity testing, p. 155-171. *In* P. Lincoln, and J. Thomsom (Eds.), Methods in Molecular Biology—Forensic DNA Profiling Protocols, Vol. 98. Humana Press, Totowa, NJ.

15. Budowle, B., F.S. Baechtel, J.B. Smerick, K.W. Presley, A.M. Giusti, G. Parsons, M. Alevy, and R. Chakraborty. 1995. D1S80 population data in African Americans, Caucasians, Southeastern Hispanics, Southwestern Hispanics, and Orientals. J. Forensic Sci. *40*:38-44.

16. Budowle, B., R. Chakraborty, A.M. Giusti, A.J. Eisenberg, and R.C. Allen. 1991. Analysis of the variable number of tandem repeats locus D1S80 by the polymerase chain reaction followed by high resolution polyacrylamide gel electrophoresis. Am. J. Hum. Genet. *48*:137-144.

17. Budowle, B., A.M. Giusti, and R.C. Allen. 1990. Analysis of PCR products (pMCT118) by polyacrylamide gel electrophoresis, p. 148-150. *In* H.F. Polesky and W.R. Mayr (Eds.), Advances in Forensic Haemogenetics 3. Springer-Verlag, Heidelberg.

18. Budowle, B., B.W. Koons, and J.D. Errera. 1996. Multiplex amplification and typing procedure for the loci D1S80 and amelogenin. J. Forensic Sci. *41*:660-663.

19. Buel, E., G. Wang, and M. Schwartz. 1995. PCR amplification of animal DNA with human X-Y amelogenin primers used in gender determination. J. Forensic Sci. *40*:641-644.

20. Cosso, S. and R. Reynolds. 1995. Validation of the AmpliFLP D1S80 amplification kit for forensic casework analysis according to TWGDAM guidelines. J. Forensic Sci. *40*:424-434.

21. Deutsch, D. 1989. Structure and function of enamel gene products. Anat. Rec. *224*:189-210.

22. Duncan, G.T., K. Balamurugan, B. Budowle, J. Smerick, and M.L. Tracy. 1996. Microvariation at the human D1S80 locus. Int. J. Legal Med. *110*:150-154.

23. Duncan, G.T., K. Balamurugan, B. Budowle, and M.L. Tracy. 1996. HinfI/Tsp509 I polymorphisms in the flanking regions of the human VNTR locus D1S80. Genet. Anal. Biomol. Eng. *13*:119-121.

24. Elliot, J.C., B. Budowle, R.A. Aubin, and R.M. Fourney. 1993. Quantitative reproduction of DNA typing minisatellites resolved on ultra-thin silver-stained polyacrylamide gels with X-ray duplicating film. BioTechniques *14*:702-704.

25. Gill, P., C.P. Kimpton, and K. Sullivan. 1992. A rapid polymerase chain reaction method for identifying fixed specimens. Electrophoresis *13*:173-175.

26. Hayes, J.M., B. Budowle, and M. Freund. 1995. Arab population data on the PCR based loci: HLA-DQA1, LDLR, GYPA, HBGG, D7S8, Gc, and D1S80. J. Forensic Sci. *40*:888-892.

27. Hjertén, S., S. Jerstedt, and A. Tiselius. 1965. Some aspects of the use of continuous and discontinuous buffer systems in polyacrylamide gel electrophoresis. Anal. Biochem. *11*:219-223.

28. Hochmeister, M.N., B. Budowle, U. Borer, and R. Dirnhofer. 1993. Effects of nonoxinol-9 on the ability to obtain DNA profiles from postcoital vaginal swabs. J. Forensic Sci. *38*:442-447.

29. Hochmeister, M.N., B. Budowle, U.V. Borer, and R. Dirnhofer. 1994. Swiss population data on the loci HLA-DQα, LDLR, GYPA, HBGG, D7S8, Gc, and D1S80. Forensic Sci. Int. *67*:175-184.

30. Hochmeister, M.N., B. Budowle, J. Jung, U.V. Borer, C.T. Comey, and R. Dirnhofer. 1991. PCR-based typing of DNA extracted from cigarette butts. Int. J. Legal Med. *104*:229-233.

31. Hochstrasser, D., A. Patchornik, and C. Merril. 1988. Development of polyacrylamide gels that improve the separation of proteins and their detection by silver staining. Anal. Biochem. *173*:412-423.

32. Honda, K., M. Nakatome, M.N. Islam, H. Bai, Y. Ogura, H. Kuroki, M. Yamazaki, M. Terada, S. Misawa, and C. Wakasugi. 1995. Detection of D1S80 (pMCT118) locus polymorphism using semi-nested polymerase chain reaction in skeletal remains. J. Forensic Sci. *40*:637-640.

33. Horn, G.T., B. Richards, and K.W. Klinger. 1989. Amplification of a highly polymorphic VNTR segment by the polymerase chain reaction. Nucleic Acids Res. *17*:2140.

34. Huang, N.E., R. Chakraborty, and B. Budowle. 1994. D1S80 allele frequencies in a Chinese population. Int. J. Legal Med. *107*:118-120.

35. Kasai, K., Y. Nakamura, and R. White. 1990. Amplification of a variable number of tandem repeats (VNTR) locus (pMCT118) by the polymerase chain reaction (PCR) and its application to forensic science. J. Forensic Sci. *35*:1196-1200.

36. Kloosterman, A.D., B. Budowle, and P. Daselaar. 1993. PCR-amplification and detection of the human D1S80 VNTR locus: Amplification conditions, population genetics, and application in forensic analysis. Int. J. Legal Med. *105*:257-264.

37. Kloosterman, A.D., P. Daselaar, B. Budowle, and E.L. Riley. 1992. Population genetic study on the HLA DQα and the D1S80 locus in Dutch Caucasians, p. 329-344. *In* Proceedings from the

Third International Symposium on Human Identification 1992. Promega Corporation, Madison, WI.

38. **Lareu, M.V., I. Munoz, M.S. Rodriguez, C. Vide, and A. Carracedo.** 1993. The Distribution of HLA DQA1 and D1S80 (PMCT118) alleles and genotypes in the population of Galicia and Central Portugal. Int. J. Legal Med. *106*:124-128.

39. **Lorente, M., J.A. Lorente, M.R. Wilson, B. Budowle, and E. Villanueva.** 1993. Composite PAGE: an alternate method for increased separation of amplified short tandem repeat alleles. Int. J. Legal Med. *106*:69-73.

40. **Ludwig, E.H., W. Friedl, and B.J. McCarthy.** 1989. High-resolution analysis of a hypervariable region in the human apolipoprotein B gene. Am. J. Hum. Genet. *48*:458-464.

41. **Mannucci, A., K.M. Sullivan, P.L. Ivanov, and P. Gill.** 1994. Forensic application of a rapid and quantitative DNA sex test by amplification of the X-Y homologous gene amelogenin. Int. J. Legal Med. *106*:190-193.

42. **Nagai, A., S. Yamada, Y. Bunai, and I. Ohya.** 1994. Analysis of the VNTR locus D1S80 in a Japanese population. Int. J. Legal Med. *106*:268-270.

43. **Nakahori, Y., K. Hamano, M. Iwaya, and Y. Nakagome.** 1991. Sex identification by polymerase chain reaction using X-Y homologous primers. Am. J. Med. Genet. *39*:472-473.

44. **Nakahori, Y., O. Takenaka, and Y. Nakagome.** 1991. A human X-Y homologous region encodes amelogenin. Genomics *9*:264-269.

45. **Nakamura, Y., M. Carlson, V. Krapcho, and R. White.** 1988. Isolation and mapping of a polymorphic DNA sequence (pMCT118) on chromosome 1p (D1S80). Nucleic Acids Res. *16*:9364.

46. **Ornstein, L. and B.J. Davis.** 1962. Disc Electrophoresis. Parts 1 and 2. Distillation Industries, Rochester, NY.

47. **Sajantila, A., B. Budowle, M. Strom, V. Johnsson, M. Lukka, L. Peltonen, and C. Ehnholm.** 1992. PCR amplficiation of alleles at the D1S80 locus: Comparison of a Finnish and a North American Caucasian population sample, and forensic case-work evaluation. Am. J. Hum. Genet. *50*:816-825.

48. **Sajantila, A., S. Puomilahti, V. Johnsson, and C. Ehnholm.** 1992. Amplification of reproducible allele markers for amplified fragment length polymorphism analysis. BioTechniques *12*:16-21.

49. **Sajantila, A., M. Strom, B. Budowle, P.J. Karhunen, and L. Peltonen.** 1991. The polymerase chain reaction and post-mortem forensic identity testing: application of amplified D1S80 and HLA-DQ alpha loci to the identification of fire victims. Forensic Sci. Int. *51*:23-34.

50. **Scholl, S., B. Budowle, K. Radecki, and M. Salvo.** 1996. Navajo, Pueblo, and Sioux population data on the loci HLA-DQA1, LDLR, GYPA, HBGG, D7S8, Gc, and D1S80. J. Forensic Sci. *41*:47-51.

51. **Skowasch, K., P. Wiegand, and B. Brinkmann.** 1992. pMCT118 (D1S80): a new allelic ladder and improved electrophoretic separation lead to the demonstration of 28 alleles. Int. J. Legal Med. *105*:165-168.

52. **Sugiyama, E., K. Honda, Y. Katsuyama, S. Uchiyama, A. Tsuchikane, M. Ota, and H. Fukushima.** 1993. Allele frequency distribution of the D1S80 (pMCT118) locus polymorphism in the Japanese population by the polymerase chain reaction. Int. J. Legal Med. *106*:111-114.

53. **Sullivan, K., A. Mannucci, C.P. Kimpton, and P. Gill.** 1993. A rapid and quantitative DNA sex test: fluorescence based PCR analysis of X-Y homologous gene amelogenin. BioTechniques *15*:636-642.

54. **Sullivan, K.M., S. Pope, P. Gill, and J.M. Robertson.** 1992. Automated DNA profiling by fluorescent labeling of PCR products. PCR Methods Appl. *2*:34-40.

55. **Thymann, M., L.J. Nellemann, G. Masumba, L. Iregns-Moeller, and N. Morling.** 1993. Analysis of the locus D1S80 by amplified fragment length polymorphism technique (AMP-FLP). Frequency distribution in Danes, intra and inter laboratory reproducibility of the technique. Forensic Sci. Int. *60*:47-56.

56. **Tully, G., K.M. Sullivan, and P. Gill.** 1993. Analysis of 6 VNTR loci by multiplex PCR and automated fluorescent detection. Hum. Genet. *92*:554-562.

57. **Woo, K.M. and B. Budowle.** 1995. Korean population data on the PCR-based loci LDLR, GYPA, HBGG, D7S8, Gc, HLA-DQA1, and D1S80. J. Forensic Sci. *40*:645-648.

58. **Yamamoto, T., C.G. Davis, M.S. Brown, W.J. Schneider, M.L. Casey, J.L. Goldstein, and D.W. Russell.** 1984. The human LDL receptor: a cysteine-rich protein with multiple Alu sequences in its mRNA. Cell *39*:27-38.

59. **Yang, F., J.L. Brune, S.L. Naylor, R.L. Apples, and K.H. Naberhaus.** 1985. Human group-specific component (Gc) is a member of the albumin family. Proc. Natl. Acad. Sci. USA *82*:7994-7998.

Short Tandem Repeat Loci Typing Using Polyacrylamide Gel Electrophoresis and Decoupled Fluorescent Scanner Detection

8

SHORT TANDEM REPEAT LOCI

A subclass of variable number tandem repeats (VNTRs) is the short tandem repeat (STR), or microsatellite, loci. The STR loci are composed of tandemly repeated sequences, each of which is 2 to 7 bp in length. Loci containing repeat sequences consisting of 4 bp (or tetranucleotides) are used routinely for human identification and, in some cases, 5 bp repeat STRs are increasingly being used. These repeat sequence loci are abundant in the human genome and are highly polymorphic (more so than the HLA-DQA1 and PM loci). The number of alleles at a tetranucleotide repeat STR locus ranges usually from 5 to 20. Because the region containing the repeats is generally quite small (a STR region with 8 copies of a 4-bp repeat is 32 bp in length), an STR locus generally is amenable to amplification by the polymerase chain reaction (PCR). Amplified alleles, in practice, are somewhat larger and can be up to 500 bases in length because the sequences that flank the repeat region where the primers reside add to the overall length. Using PCR, STRs can be typed with a high degree of specificity and sensitivity in a relatively short time period. Also, PCR multiplexing and simultaneous typing of 8 or 9 STR loci have been demonstrated, and commercial kits are available to assist in typing the multiple loci. Thus, STR typing today combines the sensitivity and specificity of PCR with a high degree of discrimination capability. The advantages of a multiplex system, as compared with analyzing each locus individually, are that less template DNA and reagents are consumed, and labor and the potential for contamination are reduced.

DNA Typing Protocols: Molecular Biology and Forensic Analysis
By B. Budowle, J. Smith, T. Moretti, J. DiZinno
©2000 Eaton Publishing, Natick, MA

STR typing has become mainstream because of the afforded sensitivity of detection and high discrimination power and because of implementation of national DNA databanks, such as the Combined DNA Index System (CODIS). The main objective for a national DNA databank is to assist investigators in the identification of perpetrators of violent crimes. For purposes of applying DNA technology to human identity testing and to make effective use of a national DNA databank, defined polymorphic genetic markers are required, and all laboratories that contribute to the database should use the same genetic loci. STR loci are the most informative PCR-based genetic markers available to date for attempting to individualize biological material. Thirteen STR loci, CSF1PO, FGA, TH01, TPOX, vWA, D3S1358, D5S818, D7S820, D8S1179, D13S317, D16S539, D18S51, and D21S11, have been selected as the core loci for use in CODIS (Table 1). A genetic profile of these STR profiles can be effective for individualization in a variety of identity testing situations.

Table 1. Thirteen CODIS STR Core Loci Characteristics

STR Name	Chromosome Location	Gene Association	Repeat Sequence Motif	9947A Type
CSF1PO	5q33.3-34	CSF-1 receptor protooncogene	AGAT	10,12
FGA	4q28	Human alpha fibrinogen	$(TTTC)_3$ TTTT TTCT $(CTTT)_n$ CTCC $(TTCC)_2$	23,24
TH01	11p15.5	Tyrosine hydroxylase	$(AATG)_n$	8,9.3
TPOX	2p23-2pter	Thyroid peroxidase	$(AATG)_n$	8,8
vWA	12p12-pter	von Willebrand antigen	TCTA $(TCTG)_{3-4}$ $(TCTA)_n$	17,18
D3S1358	3p	anonymous	TCTA $(TCTG)_{1-3}$ $(TCTA)_n$	14,14
D5S818	5q21-q31	anonymous	$(AGAT)_n$	11,11
D7S820	7q	anonymous	$(GATA)_n$	10,11
D8S1179	8	anonymous	$(TCTR)_n$	13,13
D13S317	13q22-q31	anonymous	$(GATA)_n$	11,11
D16S539	16q24-qter	anonymous	$(AGAT)_n$	11,12
D18S51	18q21.3	anonymous	$(AGAA)_n$	15,19
D21S11	21q11.2-q21	anonymous	$(TCTA)_n$ $(TCTG)_n$ $[(TCTA)_3$ TA $(TCTA)_3$ TCA $(TCTA)_2$ TCCA TA] $(TCTA)_n$	30,30

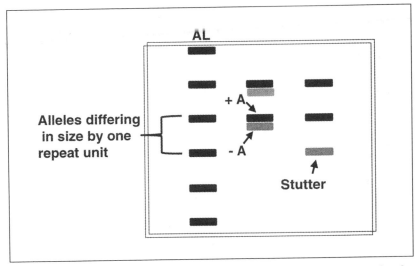

Figure 1. Schematic showing the position of +A, -A, and stutter bands. +A product (or bands) align with the allelic ladder components. -A products are one base in size smaller than the desired product. Stutter for tetranucleotide repeats tends to be one repeat (i.e., 4 bases less) smaller in size than true allelic product.

Although typically not problematic for interpretation, at times there are products that may be generated during the PCR that are not the desired allelic products. These are known as stutter bands and -A products (Figure 1).

The majority of STR loci used by the forensic community are composed of tetranulceotide repeat regions and at times exhibit stutter. The mechanism of stutter band formation may be due to slipped strand mispairing during the PCR, and stutter bands generally migrate to positions one repeat smaller than the true allele (Figures 1 and 2). The presence of stutter bands, which is substantially lower in quantity than the true alleles, does not compromise interpretation in most situations. STR loci with longer repeat regions (e.g., pentanucleotide repeats) tend to present lower stutter than tetranucleotide repeat STRs.

Taq DNA polymerase has a propensity to add a single nucleotide at the 3′ end of an amplicon in a template-independent fashion. Usually, an adenine is added (hence the term +A and - A). Other nucleotides may be added in lieu of adenine. The -A state is the true size of the target molecule, and the +A form is one base larger in size. It is easier to promote the +A state than to drive the reaction toward the -A state. Primer design and/or a final extension of 30 minutes or more at 60°C at the end of thermal cycling promotes nontemplate nucleotide addition.

The different STR alleles at a locus generally are based on the number of tetranucleotide repeats that they contain. However, some alleles can contain partial repeats (i.e., less than 4 bases) within the array of tetranucleotide repeats. Designation of these alleles is straightforward (Figure 4). For example, the TH01 9.3 allele is a relatively common allele in Caucasians that is one base pair smaller in size than the 10 allele. It contains 9 complete tetranucleotide repeats and a partial repeat with only 3 bases.

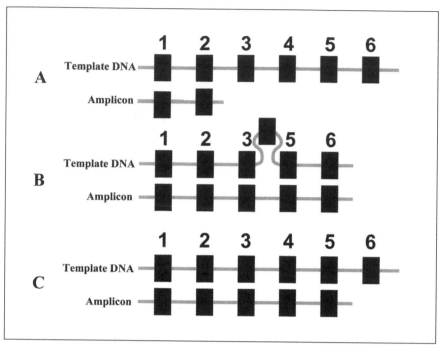

Figure 2. A proposed mechanism for stutter product formation is strand slippage. A = primer extension is occurring through the repeat region; B = because of processivity of *Taq* DNA polymerase, the copying process stops. The duplex relaxes allowing for the template to loop-out; C = when elongation continues and the duplex is stable, the amplicon and template are out of phase by a repeat unit. Thus, a stutter product is one repeat smaller in size than the true allele.

The process for typing the amplified STRs in this chapter entails separating the fragments, usually by polyacrylamide gel electrophoresis, and detecting the products after separation has been completed. The electrophoretic gel contains a denaturant so that the amplicons are separated as single-stranded molecules. Better separation of the STR alleles is achieved using denaturing gel electrophoresis (as is done for sequencing described in Chapter 10, entitled PCR-Based Analyses: Mitochondrial DNA Sequencing) than when employing native (or nondenaturing) gel electrophoresis. Also, because of the lower temperature of native gel electrophoresis, the mobility of double-stranded DNA may alter its mobility due to conformational changes. Most separations of STR fragments are carried out under denaturing conditions to avoid complications in typing alleles that may occur due to conformational changes.

After electrophoresis, the separated amplicons can be stained using a general stain, such as silver (as described in the previous chapter) or by labeling the primers with a fluorescent tag (so that the tag will be incorporated into the amplicons during the PCR). Fluorescently-labeled STR amplicons can be analyzed in a decoupled fashion using a fluorescent scanner. Decoupled means that electrophoresis is completed prior to scanning for fluorescently labeled amplicons. This detection platform is equipped with a laser, filters, and an emission detection device. After electrophoresis, the gel is removed

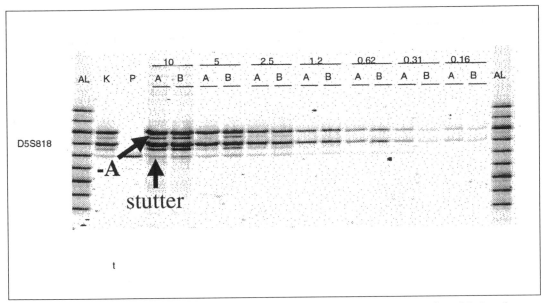

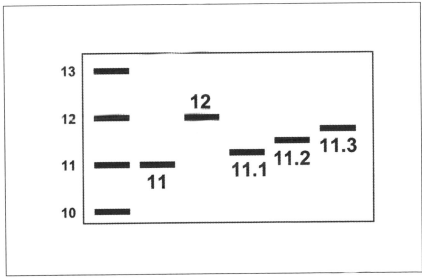

Figure 3. Position of -A and stutter for STR locus. The locus is D5S818. Decreasing amounts of amplified product are shown. The more overloaded samples tend to show -A or stutter. This polyacrylamide gel was scanned on a FMBIO II fluorescent scanner.

Figure 4. Allelic designations for off-ladder variants of STR loci. Alleles are designated by comparison to allelic ladders that contain alleles of known size (based on number of repeats). Generally, the repeats are composed of four bases each. The numbers to the left represent the number of repeats contained in each allele of the ladder. Those alleles that do not line up with the allelic ladder alleles (known as off-ladder alleles) are designated by the total number of repeats (whole number before decimal point) plus the number of nucleotides comprising the partial repeat unit (after decimal point).

from the electrophoresis apparatus and subsequently scanned using a fluorescent scanner. Electrophoretic profiles can be stored electronically. Real time detection of STR alleles can also be achieved during electrophoresis and will be discussed in Chapter 9, entitled PCR-Based Analyses: Capillary Electrophoresis for STR Loci Typing and Real Time Fluorescent Detection.

The DNA samples can be amplified simultaneously at the loci CSF1PO, TPOX, TH01, VWA, D5S818, D7S820, D13S317, and D16S539 using the *GenePrint*™ PowerPlex™ 1.1 System (Promega, Madison, WI, USA) (i.e., PowerPlex kit) and a GeneAmp® PCR System 9600 DNA Thermal Cycler (PE Biosystems, Foster City, CA, USA). The GenePrint PowerPlex 1.1 System contains all reagents for the PCR except the Taq DNA polymerase. *Taq* or Ampli-*Taq* Gold™ (PE Biosystems) may be used in the PCR. One of the primers for each of the loci D5S818, D7S820, D13S317, and D16S539 is labeled with fluorescein, and for the loci CSF1PO, TPOX, TH01, and VWA one primer for locus is labeled with carboxy-tetramethylrhodamine. The GenePrint PowerPlex 2.1 System enables simultaneous amplification of 9 STR loci. One of the primers for each of the loci Penta E (a pentanucleotide repeat locus), D18S51, D21S11, TH01, and D3S1358, is labeled with fluorescein, and for the loci FGA, TPOX, D8S1179, and VWA the primer is labeled with carboxy-tetramethylrhodamine. Thus, the 13 core STR loci for CODIS can be amplified using the GenePrint PowerPlex 1.1 and GenePrint PowerPlex 2.1 Systems. Multicolor detection enables an increase in the number of loci that can be analyzed simultaneously. Loci of similar size (that superimpose each other) can be resolved if labeled with different colored fluors, if the scanning/detector device is capable of separating the fluor emissions. These fluors are compatible with the FMBIO II fluorescent scanner (Hitachi Genetic Systems/MiraiBio, Alameda, CA, USA), which is used to detect the separated amplicons. The preparation and running of the gels for silver staining or the decoupled fluorescent scanning method is similar. The formulae to calculate the polyacrylamide gel percentage and percent cross-linker are provided to facilitate the user to prepare appropriate gels as needed.

8.1 AMPLIFICATION

Materials and Reagents

- GeneAmp PCR System 9600 DNA thermal cycler
- *GenePrint* PowerPlex 1.1 System
- *GenePrint* PowerPlex 2.1 System
- STR 10× buffer or GoldSTR buffer
- PowerPlex 10× primer pair mix
- PowerPlex allelic ladder
- Internal lane standard 600 or CXR fluorescent ladder 60 to 400 bases
- K562 cell line DNA
- Bromophenol blue loading solution
- Gel tracking dye
- Aerosol resistant tips

Procedure

1. Thaw STR 10× buffer and PowerPlex (1.1 or 2.1 System) 10× primer pair mix and place on ice. Prior to use, mix reagents by vortex-mixing.

2. Create sufficient PCR master mixture for all samples to be assayed (allow for 1 or 2 extra reactions to compensate for measurement error during pipetting).

3. Label and place 0.2- or 0.5-mL microcentrifuge tubes in rack.

4. In the following order [17.05 µL water, 2.50 µL 10× buffer, 2.50 µL primer mix, 0.45 µL polymerase (2.25 U)] add the calculated master mixture volumes to a 1.5-mL tube. The amount of nuclease-free water in the PCR should be adjusted accordingly so that the final volume of master mixture per each reaction is 22.5 µL.

5. Add 22.5 µL PCR master mixture to each PCR tube on ice.

6. Add 1–2 ng of template DNA from samples or control (in 2.5 µL total volume)

7. For the negative control, add 2.5 µL of nuclease-free water.

   Note: No addition of mineral oil is required.

8. Briefly centrifuge the tubes to bring contents to the bottom of each tube.

9. Place tubes in the thermal cycler.

10. Select and run a recommended thermal protocol

    Cycling profile: 96°C 2 min incubation/initial denaturation

    > 94°C 30 s
    > Ramp 68 s to 60°C, hold for 30 s
    > Ramp 50 s to 70°C, hold for 45 s
    > For 10 cycles, then

    > 90°C 30 s
    > Ramp 60 s to 60°C, hold for 30 s
    > Ramp 50 s to 70°C, hold for 45 s
    > For 20 cycles

    > 60°C 30 min

    Note: If using AmpliTaq Gold, an initial incubation of 11 min at 95°C is required, instead of 2 min.

    Note: The final 30-min extension at 60°C is to drive all products to be adenylated (i.e., +A).

11. After PCR, store samples at -20°C.

8.2 POLYACRYLAMIDE GEL PREPARATION

Materials and Reagents

- Model SA 32 Gel Electrophoresis System (Life Technologies, Gaithersburg, MD, USA)

- Acrylamide
- Bisacrylamide
- 10% (wt/vol) ammonium persulfate
- Tetramethylethylenediamine (TEMED)
- 10× Tris-borate-EDTA (TBE) (1.0 M Tris, pH 8.4, 0.9 M boric acid, and 0.01 M EDTA)
- Glass plates
- Sample loading buffer (10 mM NaOH, 95% formamide, 0.05% bromophenol blue, and 0.05% xylene cyanol FF)
- Urea
- Bind silane (1.5 µL silane to 500 µL of 95% ethanol/0.5% acetic acid)

Preparing a Gel

Procedure

a. Gel stock solution:
   30% monomer (T), 5% cross-linker (C) (28.5 g acrylamide, 1.5 g bisacrylamide/100 mL)

b. Formulae for calculating %T and %C:
   %T   = a+b/m
   %C   = b/a+b
   (a = g of acrylamide; b = g of bisacrylamide; m = total gel volume)

c. Gel recipe (6.5% T, 5% C, 1× TBE):
   gel stock solution   = 6.5 mL
   10× TBE   = 3.0 mL
   urea   = 13.5 g [7.5 M final concentration—be certain that urea is completely in solution and then filter entire solution through a 0.2-µm (Nalgene) vacuum filter.]

   Bring to a final volume of 30 mL with double distilled water and mix by swirling gently.

d. When ready to pour the gel add 10% ammonium persulfate, 200 µL, and TEMED, 20 µL.

Note: Polymerization begins upon addition of the ammonium persulfate and TEMED, which is a catalyst; you must instill the gel into the cassette immediately.

Gel Setup

1. Use low fluorescence glass plates if fluorescent detection is desired. (Silver staining does not require low fluorescence glass plates.) Prior to each use, the glass plates used for gel preparation must be cleaned thoroughly. Wash each plate and then clean with water and/or methanol. Wipe dry with a lint-free laboratory wiper towel. Wash the Delrin comb with water and allow it to dry.

2. Treat the long plate with gel slick and treat the short plate with bind silane.

3. To assemble the plates for use, place the long plate on a flat surface with the long axis of the plate parallel to the edge of the laboratory bench.

4. Place 0.4-mm plastic spacers along each side of the plate. The ends of the spacers should be even with the end of the plate. In addition, the ends of the spacers with the foam cushions should be opposite the end of the plate. Place the short plate on top of the spacers, with the foam cushions pressed tightly against the top of the short plate. Make certain the spacers are even with the edges of the plates. Place three binder clips on each side of the gel cassette (with the pressure of the clips exerted on the spacers). Place the gel cassette on a level surface.

5. Draw the gel solution into a syringe or pipet, taking care not to introduce any air bubbles, and inject the solution into the cassette from the top of the short plate with the cassette slightly tilted. Continue to instill the gel solution until the injection front reaches the bottom of the cassette. The length of time for injection of the gel solution should be no more than 60 seconds.

6. Immediately insert the comb into the gel and press it tightly against the top edge of the short plate. Allow the gel to polymerize for at least 1 h at room temperature. If the gel is to be stored, cover with plastic wrap to prevent drying and refrigerate at 4°C.

8.3 SAMPLE PREPARATION FOR ELECTROPHORESIS

Procedure

1. Mix 3 μL of loading buffer with 3 μL of PCR product (an internal standard can be added, if desired).

2. Denature samples at 95°C for 2 min and place on ice.

3. Load 5 μL onto the cathodal end of the gel.

    Note: These quantities are provided for instructional guidance; they can and will change depending on the PCR method/kit employed.

8.4 ELECTROPHORESIS

Procedure

1. Place gel/plate sandwich on the Model SA 32 electrophoretic apparatus.

2. Add 1× TBE buffer to the bottom reservoir of the electrophoretic apparatus. Install the gel cassette on the electrophoretic apparatus with the long plate to the outside. Tighten the gel clamps firmly and make certain that the drain clamp is closed. Remove any bubbles that may have been trapped at the bottom of the gel. Add 1× TBE buffer to the top tank of the electrophoretic apparatus. Carefully remove the comb from the gel.

3. Using a syringe, flush out the wells in the gel by repetitively forcing buffer into the wells.

4. Using special pipet tips (UniFit disposable, flat microflex tips), load a 5-μL sample volume into a well of the gel. The sample volume will be loaded underneath the TBE buffer that will be in the wells. Slowly pipet the sample into the wells so as to form a smooth layer under the buffer.

5. Close the tank covers and attach the electrode wires (cathode at the top). Set the power supply to deliver 80 W. Set the milliamperage and voltage to maximum. After 5 min, reset the wattage to 40.

6. Allow the electrophoresis to proceed until the xylene cyanol dye marker (turquoise) reaches the bottom of the plates (about 2–2.5 hours). Stop the run, drain the buffer from the top tank, and remove the gel from the electrophoresis unit. Do not separate the glass plates. The gel is now ready for scanning.

   Note: For some STR multiplex systems, the alleles of the largest STR loci (those residing at the top of a gel near the sample origin) will not be separated as well as the alleles for smaller sized STR loci. The glass plates, which still house the gel, can be placed back intact on the electrophoresis apparatus. Place 1× TBE buffer in the upper reservoir and continue electrophoresis at 40 W for 1 h (the actual time may vary depending on the size of the STR alleles. Alternatively, longer gels can be run using a Model SA 43 apparatus (Life Technologies).

8.5 POLYACRYLAMIDE GEL ANALYSIS USING THE FMBIO II

The FMBIO II fluorescent scanner (Hitachi Genetic Systems/MiraiBio; Alameda, CA, USA) can accommodate four filters. The 605-nm filter must be in Channel 1 to enable the Autofocus function. The other filters can be placed as desired in the other Channels. These instructions are based on the following filter configuration: Channel 2, 585 nm; Channel 3, 505 nm; and Channel 4, 650 nm.

Scanning a Gel

Procedure

1. Turn on power to the computer and the FMBIO II (the switch is behind the pull-down panel on the front left).

2. To prepare the gel for scanning, carefully clean and dry the outside of the glass plates. Do not take the gel plate assembly apart.

3. Place the gel plate assembly in the scanner.

   a. Press the button on the front of the instrument to open the sample door.

   b. Slide the right side stage bar to the proper distance to support the plate, without obscuring the area to be scanned.

   c. Place the plate assembly (long plate on bottom) on the stage, centering the plate assembly on the black bars at each end. Make sure the plate is level and secure.

4. Start the FMBIO II ReadImage program (for MacIntosh®).

5. Select NEW (from dialog box or file menu) to set the parameters. The ReadImage Parameter dialog box will appear.

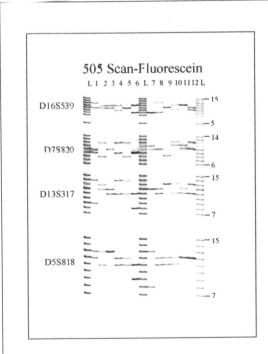
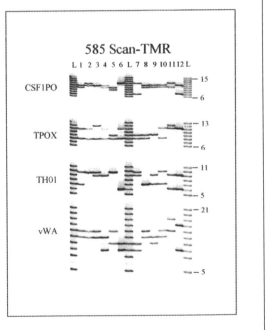

Figure 5. *GenePrint* **PowerPlex 1.1 System displaying 8 STR loci. The samples were subjected to polyacrylamide gel electrophoresis on an SA 32 apparatus and scanned on a FMBIO II fluorescent scanner (figure kindly provided by Promega, Madison, WI, USA).**

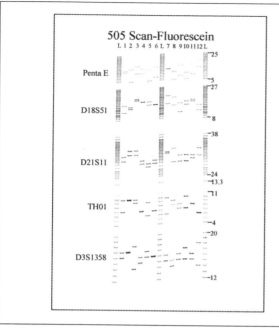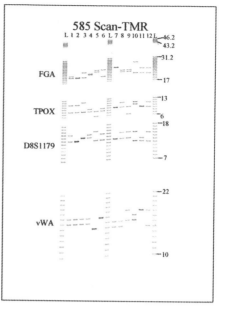

Figure 6. *GenePrint* PowerPlex 2.1 System displaying 9 STR loci. The samples were subjected to polyacrylamide gel electrophoresis on an SA 32 apparatus and scanned on a FMBIO II fluorescent scanner (figure kindly provided by Promega).

6. Set the Scanning parameters as follows:

   Material Type: acrylamide gel
   Resolution: 150 dpi × 150 dpi
   Repeat: 256
   Sensitivity: 1CH = 100%, 2CH = 80%, 3CH = 100%, 4CH = 100%
   Focus: 0 (polyacrylamide gel with 5-mm plates)
   Note: The focus value may need to be lower for the thinner plates used in precast gels. The range of focusing is -2 to +3 mm. Use of values outside of this range can seriously damage the instrument.

7. Click Ok, and the ReadImage main screen appears.

8. In the FMBIO II menu, check channels 2, 3, and 4 to enable collection of data through the appropriate filters. Channels 2, 3, and 4 will be listed as the Active Channels in the ReadImage main screen.

9. Since the instrument always autofocuses before the first read after power-up, turn Autofocus off in the FMBIO II menu.

10. In the COMMAND menu, set the Gray Level Correction as follows:

    Low: 50% (Background)
    High: 1% (Signal)

11. To save the ReadImage parameters for future scans, choose SAVE from the FILE menu, assign a filename and location, and then click SAVE. Double-click on the saved parameter file, launch the ReadImage program, and return the user to the saved scanning parameters.

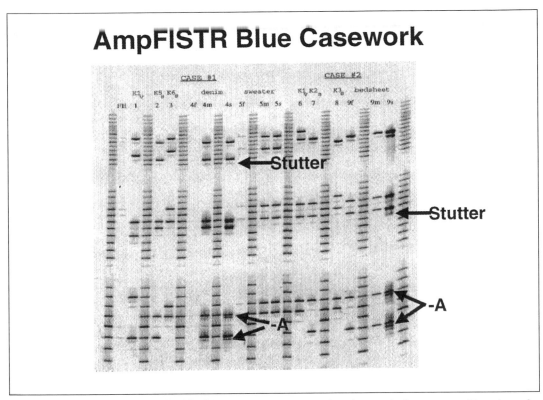

Figure 7. DNA from casework samples. The DNA was amplified using a three-locus multiplex system, subjected to poly-acrylamide gel electrophoresis on an SA 32 apparatus, and scanned on a FMBIO II fluorescent scanner.

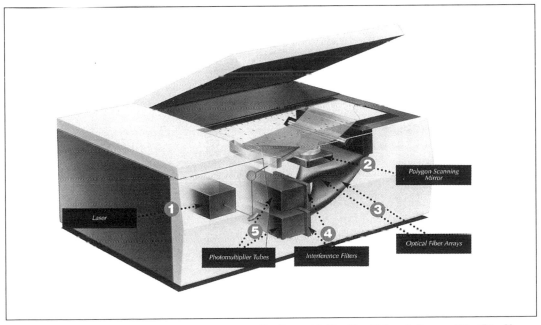

Figure 8. Schematic of the FMBIO II fluorescent scanner (kindly provided by Hitachi Genetic Systems/MiraiBio, Alameda, CA, USA).

Note: If you do not want to see the directory dialog when you start Read-Image, open EDIT > PREFERENCES and toggle off Show open dialog at start up. The Scan Control dialog box will then open when ReadImage is launched.

12. The Scan Control dialog box appears. Comments associated with the image file may be typed in the Comment field at the bottom of the Scan Control dialog box.

13. Choose PREREAD to quickly identify the gel area for the final scan.

Note: If a small scanning box is already set from a previous scan, click the ALL AREA button before prereading. If you need to terminate the preread in progress, choose PAUSE or CANCEL.

14. Draw a box around the desired scan area by dragging with the mouse from the top left to the lower right of the desired scan area.

Note: Do not include spacers in this box because they can distort the grayscale balance of the image.

15. Choose READ. Specify a file name (maximum of 20 characters) and a location for the file to be saved. Images from scans using more than one channel will be placed automatically together within their own folder at the specified location.

Note: You can terminate a scan before completion by choosing PAUSE and CANCEL. You can then save or discard the image.

16. Wait for the instrument to complete all of its scans. Depending on the channels activated, the scanner may pass over the gel twice.

8.6 MULTICOLOR GEL ANALYSIS USING FMBIO ANALYSIS SOFTWARE VERSION 8.0

Procedure

1. Open the FMBIO ANALYSIS program (version 8.0 for MacIntosh).

2. Choose FILE > NEW. A directory dialog opens with the message SELECT MULTI ANALYZE IMAGE. Select an image file in the directory window (any channel) and choose OPEN.

Note: If a multicolor project has already been saved, you can open it using the FILE > OPEN command.

3. A directory dialog opens with the message SAVE NEW PROJECT AS. Enter a file name and choose SAVE.

If the gel has not been analyzed before, a MARKER MODE window may appear. If you are using an Internal Lane Standard [i.e., carboxy-X-rhodamine (CXR)-labeled], choose LAYER.

Note: In the pull-down box to the right of LAYER, select the channel containing the Internal Lane Standard (i.e., 4CH, for the CXR label).

If you are not using an Internal Lane Standard, choose SEPARATE.

An image window opens to display the multicolor image. The MULTI menu appears in the FMBIO ANALYSIS menu bar.

The PROJECT palette also appears and stays open as long as the project is open. You can click the PROJECT palette and move it to the edge of the screen to have an unobstructed view of the Image window.

To make a particular color invisible on the gel image, click the black dot above the color's box in the PROJECT palette.

To change the intensity of each image color, click and move the sliding Color Intensity bar in the PROJECT palette to the right or left.

Note: If the scanned area was larger than desired, you can select the region of interest and save the smaller image. Use the MARQUEE tool to draw a box around the desired region of the image, and choose FILE > SAVE SPECIAL > SELECTION.

4. To perform grayscale adjustment, choose MULTI > DISPLAY MODE > MONO, or click the MONO MODE tool in the PROJECT palette. Image is converted to grayscale. It may be helpful to increase the magnification of the image before adjusting the grayscale. In the lower left tool box, click the ENLARGE tool one time to reach 200 (200 will be indicated just above the ENLARGE tool). After enlargement, click the FINE PIC-TURE button (eye symbol) to improve the appearance of the enlarged image.

5. Do the following for each channel successively:

   a. Select an image layer by clicking the appropriate channel NAME in the PROJECT palette.

   b. In the lower left tool box, click the GRAY LEVEL ADJUSTMENT tool.

   c. A GRAY LEVEL ADJUSTMENT dialog box appears. Place the window of this box over the region of interest by clicking and dragging the frame of the window to the desired location.

Table 2. The Recommended Multi-Band Separation Parameter Settings		
	PowerPlex 1.1	**PowerPlex 1.2**
Overlap:	0.0 mm	0.0 mm
Band Area:	0.1 mm	1.0 mm
Background Area 1&2:	1.0 mm	1.0 mm
Background:	100%	100%
Note: These parameters can be saved by clicking the SAVE button and assigning a unique name.		

d. Click the LOW (BACKGROUND) % or VALUE field. Place the cursor in the viewing window and over a region of background signal. Use the mouse to draw a box around a region of the image that represents the typical background. The region of the gel within the viewing window will reflect the grayscale adjustment. If the result is not satisfactory, draw another box around a background region to repeat. To return to the unaltered image, choose ORIGINAL.

e. Click the HIGH (SIGNAL) % or VALUE field, and draw a very small box within the center of a band that is of moderate intensity. The grayscale adjustment will be reflected within the viewing window. If the result is not satisfactory, select another band to designate as the high signal. When the result is satisfactory, click SET.

f. To test the gray level adjustment in other regions of the image, move the GRAY LEVEL ADJUSTMENT dialog box to another area and click TRY. You can move the dialog box and click TRY repeatedly. If necessary, you can continue to modify the gray level to improve image clarity.

Repeat steps a through f for the other channels.

Note: If the gray level settings are to be used for future analyses, select save, and assign a unique name.

FMBIO Analysis recalculates the gray levels for the entire image using the adjusted values as high and low limits. The raw image data will not be affected by gray level settings.

6. In the PROJECT palette, click the BLEND tool (red, yellow, green) to return to the multicolor image.

Click the REDUCE tool to return the magnification to 100.

7. Choose TOOLS > 1D-GEL. The 1D-Gel tools now appear to the left of the gel image, and 1D-GEL now appears in the upper menu bar.

Choose 1D-GEL > SETTING > MARKER MODE.
Make sure that the layer with the internal size standard is chosen (i.e., 4CH) and that layer mode is checked. Click OK.

8. To perform color separation, first review the Multi-Band Separation Parameters.
Choose 1D-GEL > SETTING > SEPARATION MARKER.
The MULTI-BAND SEPARATION PARAMETERS window opens.

a. In the 1D-GEL tool box, click the SINGLE LANE tool. With this tool, click the center of a lane at its top and drag the icon down the entire length of the lane. The chosen lane should have at least one nonoverlapping band of each color with a maximum of 40 bands. When you release the mouse button, move the mouse laterally to define the width of the lane and then click the mouse button again.

Note: When a lane is selected, the blue lane lines are stippled and vibrating. When a lane is not selected, the blue lanes are solid. You can select a lane by clicking the middle blue line for the desired lane. You can select multiple lanes for latter functions by holding down the SHIFT key and clicking the middle line of multiple lanes successively. In most functions, to select all lanes, you can click APPLE A, or click EDIT > SELECT ALL.

b. With the lane highlighted, choose the AUTOBAND button in the 1-D GEL tool box. The software will find and mark the bands with a horizontal line.

c. Choose MULTI > COLOR SEPARATION. The COLOR SEPARATOR window appears.

Note: The following steps must be done in the order indicated, without going back or doing any other steps.

d. With the lane highlighted, click the MULTI button in the COLOR SEPARATOR window. The MULTI-BAND COLOR SEPARATOR window appears, with a color assigned to each nonoverlapping band.

Note: In some cases, there will be fewer bands displayed in the Multi-Band Color Separator than were defined by the AutoBand function. Only bands that do not overlap are used for multi-band color separation.

Typically, the Multi-Band Color Separator accurately selects bands and assigns colors to them. In rare instances, you may want to edit the bands displayed. If a band was not assigned the appropriate color, then the color can be changed by clicking on the colored band of interest in the right hand columns and selecting the appropriate color. If a selected band is not to be used, click on the check-mark in the USE box to deselect the band.

e. Choose >>COPY>>. You will be asked, COPY SELECTED LANE INFORMATION TO COLOR USE? Choose OK.

f. Click the APPLY button. You will be asked, APPLY TO COLOR SEPARATION PARAMETER? Choose OK.

g. The COLOR SEPARATOR box appears. Click the MATRIX button. The MATRIX FOR COLOR SEPARATION box appears. If the matrix is to be used in the future, click SAVE AS and assign a unique name for the matrix.

h. Click CLOSE. In the COLOR SEPARATOR window, click OK. You will be asked, DO YOU WANT TO EXECUTE COLOR SEPARATION? Choose OK. The resulting image should display only the signal from a single dye in each channel.

9. To set the lanes, first delete the lane lines you drew on one lane for color separation. This is done by highlighting the lane and hitting the DELETE key, or by selecting from the upper tool bar EDIT > CLEAR. Then, in the 1D-GEL tool box, select the MULTIPLE LANES tool.

Move the cursor to the center of the leftmost lane. Click and hold the mouse button while dragging the mouse to the center of the rightmost lane. Release the mouse button and drag the mouse down the lanes to the bottom of the gel. Click the mouse button again.

10. The MULTI LANES window will appear. Enter the number of lanes in the gel, select the DO LANE FITTING box, and choose OK. Blue lines appear on all the lanes. You can edit them by selecting the arrow and:

Table 3. The AUTOBAND Settings		
	PowerPlex 1.1	**PowerPlex 2.1**
STARTING SLOPE:	2.5	1.5
ENDING SLOPE:	2.5	1.5
DURATION:	0.2	0.1
NOISE LEVEL:	20	20

a. To shift the entire lane line left/right, click the center line of a high-lighted lane and move the mouse as desired.

b. To angle the lane for curved lanes, click the middle blue square at the bottom of a highlighted lane line and drag the mouse as desired.

c. To change the width of a lane, click the left or right blue square at the bottom of highlighted lane and drag the mouse as desired.

Note: The software does not assign lane numbers to blank lanes and negative control lanes (if blank). If there are such lanes, you will see a window asking you to verify lane placement of blank lanes.

Note: If this lane configuration is to be used in the future, select MAKE LANE TEMPLATE from the 1D-GEL menu and assign a unique name.

11. If the white migration lines are not seen at the top and bottom of the gel, select the SHOW RANGE tool in the 1D-GEL tool box.

Move these lines up or down as desired by clicking on the white square and dragging the mouse.

Note: For ease of marker assignment, laboratories should standardize their placement of the top migration line relative to the internal lane standards.

12. With the lanes selected and the migration lines adjusted, autoband determination can be performed.

Click 1D-GEL and make sure VOL CALC is checked.
Click 1D-GEL > SETTING > CALCULATION and make sure TYPE 4 and OD CALC WITH BACKGROUND are checked.

13. Check autoband parameters. Choose 1D-GEL > SETTING > AUTO-BAND.

A SET AUTOBAND window appears.
Choose OK.

14. Highlight all lanes (press Apple A).

a. In the 1D-GEL tool box, click the AUTOBAND button.

b. Click the MONO MODE button in the PROJECT window.

c. The user may need to adjust the automatic band calls made by the software. To do so, first click on the NAME of a channel in the PRO JECT palette. Then:

In the ID-GEL tool box, click the LANE MARKER tool until the blue lines disappear, and click the SHOW BANDS tool until the yellow band area lines disappear and only the red band marker lines are shown.

To see underneath the band lines, click on the image and hold the mouse button down.

To delete a band mark, select the MARQUEE tool and draw a box around the undesired band mark. Press the DELETE key. To add a band line for a band not found by the software, click on (blacken) the PEAK SEEKER tool (it is advisable to use the PEAK SEEKER to automatically place the band mark on the area of highest signal nearest to where the user identifies a band). Click the MANUAL BANDING tool and click on the band that you want to mark.

d. Repeat step c for the other channels.

15. To set the marker (internal lane standard) in the PROJECT window, click on the NAME for 4CH.

a. In the 1D-GEL tool box, click SET MARK. The SETTING FOR STANDARD MARKER ANALYSIS window appears. Select NATURAL LOGARITHMS and click OK.

b. The user will be asked to assign a marker to the lanes on the gel. Ensure that the layer shown on top of the marker window is the layer in which the internal lane standards were run. If it is not, select the correct layer from 1D-GEL > SETTING > MARKER.

In the SETTING MARKER box, select all lanes in the box that contain the internal lane standard by holding down the SHIFT key and clicking on each lane. In the pull-down menu, select the appropriate marker (i.e., MCXR600 for PowerPlex 1.2). Click OK to apply the selected marker to the lanes on the gel.

16. To calculate base pairs, in the 1D-GEL tool box, click the CALCULATOR tool.

Results can be viewed by clicking the TABLE button in the 1D-GEL tool box. Switch between layers using the PROJECT palette.

17. The size data must be saved in DAT format to enable the files to be exported to STaR Call:

In the PROJECT palette, click the NAME of a channel, and click the TABLE button.
Choose FILE > SAVE ANALYSIS.
It is recommended that you keep the name as is for easy retrieval by STaR CALL.
The file should be kept in a folder specific for a single gel. The file extension must be .DAT.

Note: Do not modify (reformat) this table if you plan to merge the data from different channels into a single file (see section entitled Designating Alleles Using STaR Call).

In the PROJECT palette, click the NAME of another channel and repeat step 16 for each channel (saving each CH.DAT separately).

18. Close out the tables and click on the image. Click FILE > SAVE to save any modifications you made to the gel image.

19. Quit FMBIO Analysis.

8.7 DESIGNATING ALLELES USING STaR CALL

Procedure

1. Open the STaR CALL program.
   The Excel table CODISDBF.XLS appears. Do not close or modify this table because STaR Call uses it for allele calling.

2. Choose STaR CALL > IMPORT STR/RFLP. A FILE SELECTION window opens. Select a DAT file and click OPEN.

   Note: Always use the IMPORT command to open an FMBIO Analysis results file in STaR Call. Errors occur when you use the OPEN command to open a saved results file.

   The IMPORT STR/RFLP function converts an FMBIO-generated table to a STaR Call table format. It removes all unnecessary columns (such as the mm and Rf columns) and adds columns for allele scoring. It also adds rows for CODIS export information.

3. A CONVERT TABLE AS window appears. Choose STR WITH OD VALUES.

4. Click STaR CALL > EVALUATE STR DATA.

   A dialog box appears, displaying lanes and the number of bands in each lane. You are prompted to choose the lanes that contain allelic ladders. The default selection is all of the lanes that contain the highest number of bands. Select multiple lanes containing allelic ladders by pressing the COMMAND key as you click the lanes.
   Click OK to complete allelic ladder selection.

   Note: This is a good time to verify that the correct number of bands are in the chosen lanes. The number in parentheses preceding the lane number indicates the number of bands in the lane.

5. A SELECT LOOKUP TABLE dialog box appears.
   Make the appropriate selection for the channel to be analyzed.
   Click OK.

6. SAVE these results as an EXCEL file. In the pull-down box, if EXCEL is not set as the default, you will need to click SAVE FILE TYPE AS > MICROSOFT EXCEL WORKBOOK. You may keep the same file name (saved previously as .DAT) and replace the DAT file with the EXCEL file.

A SUMMARY INFO box appears. Enter information if desired. Close file.

7. Repeat steps 2 through 6 for the other channel that contains STR alleles.

8. The data from the two channels (i.e., fluorescein-labeled alleles and tetramethyl-rhodamine-labeled alleles) can be merged into a single table that can be exported to CODIS.

    Close all DAT files. Click STaR CALL > MERGE.
    The MERGE DAT FILES dialog box appears. Click the top SELECT button and select a DAT file for one channel.
    Click the next SELECT button and select the DAT file for the second channel. Click OK.

9. Click STaR CALL > VIEW STR (LANDSCAPE). The merged file is created. Print and save results.

SUGGESTED READING

1. Budowle, B., B.W. Koons, K.M. Keys, and J.B. Smerick. 1996. Methods for typing the STR triplex CSF1PO, TPOX, and HUMTHO1 that enable compatibility among DNA typing laboratories, p. 107-114. *In* A. Carracedo, B. Brinkmann, and W. Bar (Eds.), Advances in Forensic Haemogenetics 6. Springer-Verlag, Berlin.

2. Budowle, B., T. Moretti, K. Keys, B. Koons, and J. Smerick. 1997. Validation studies of the CTT multiplex system. J. Forensic Sci. *42*:701-707.

3. Crouse, C.A. and J. Schumm. 1995. Investigation of species specificity using nine PCR-based human STR systems. J. Forensic Sci. *40*:952-956.

4. Edwards, A., A. Civitello, H. Hammond, and C.T. Caskey. 1991. DNA typing and genetic mapping with trimeric and tetrameric tandem repeats. Am. J. Hum. Genet. *49*:746-756.

5. Edwards, A., H.A. Hammond, L. Jin, C.T. Caskey, and R. Chakraborty. 1992. Genetic variation at five trimeric and tetrameric repeat loci in four human population groups. Genomics *12*:241-253.

6. Kline, M., D. Duewer, P. Newall, J. Redman, D. Reeder, and M. Richard. 1997. Interlaboratory evaluation of short tandem repeat triplex CTT. J. Forensic Sci. *42*:897-906.

7. Polymeropoulos, M.H., D.S. Rath, H. Xiao, and C.R. Merrill. 1991. Tetranucleotide repeat polymorphism at the human tyrosine hydroxylase gene (TH). Nucleic Acids Res. *19*:3753.

8. Promega. 1999. *GenePrint*™ PowerPlex™ 1.1 System. Technical Manual. Part No. TMD008. Madison, WI.

9. Promega. 1999. *GenePrint*™ PowerPlex™ 1.2 System. Technical Manual. Part No. TMD011. Madison, WI.

10. Puers, C., H. Hammond, L. Jin, C. Caskey, and J. Schumm. 1993. Identification of repeat sequence heterogeneity at the polymorphic short tandem repeat locus HUMTHO1 [AATG]n and reassignment of alleles in population analysis by using a locus-specific allelic ladder. Am. J. Hum. Genet. *53*:953-958.

11. Van Oorschot, R., S. Gutowski, S. Robinson, J. Hedley, and I. Andrew. 1996. HUMTHO1 validation studies: effect of substrate, environment, and mixtures. J. Forensic Sci. *41*:142-145.

Capillary Electrophoresis for Short Tandem Repeat Loci Typing and Real Time Fluorescent Detection

9

CAPILLARY ELECTROPHORESIS

In Chapter 8, entitled PCR-Based Analyses: Short Tandem Repeat Loci Typing Using Polyacrylamide Gel Electrophoresis and Decoupled Fluorescent Scanner Detection, electrophoresis of short tandem repeat (STR) fragments was described as being carried out in polyacrylamide slab gels. However, capillary electrophoresis (CE) provides an alternative analytical tool for the separation of DNA fragments. The use of a capillary rather than a slab gel for separation of DNA fragments can augment automation, thus reducing manual efforts of the analyst and reducing reliance on the manual skills of the analyst. While many of the same parameters affecting conventional electrophoretic systems [such as electroendosmosis (EEO), peak/band diffusion, joule heating, electrodispersion, interaction with the medium and surfaces, and electrophoretic separation conditions] are relevant to CE, the process has a number of desirable features. With CE, manual gel pouring and sample loading are eliminated; loading a sieving medium into the capillary and sample injection are achieved by automatic means. The high surface area to volume ratio of a capillary enables electrophoretic separations to be carried out at high field strengths. Faster separation times are thereby achieved, and resolution may be improved compared with some slab gel electrophoresis methods. Furthermore, real time detection is performed with CE. Results are stored directly in the computer, thus facilitating subsequent data manipulations and analyses.

DNA Typing Protocols: Molecular Biology and Forensic Analysis
By B. Budowle, J. Smith, T. Moretti, J. DiZinno
©2000 Eaton Publishing, Natick, MA

A CE apparatus typically consists of a fused silica capillary positioned between two electrolyte reservoirs. In each buffer reservoir resides a high voltage electrode that is connected to a DC high voltage power supply. A detection window in the capillary, for monitoring migration of DNA fragments, is located distal to the sample injection end of the capillary. CE generally entails: *(i)* preparation of the capillary, *(ii)* loading the capillary with a separation medium and buffer, *(iii)* injection of the sample, *(iv)* separation of the sample components, *(v)* detection of the separated species, and *(vi)* subsequent data analyses.

Because the charge-to-mass ratio of different size DNA molecules is the same, mobility in free solution is independent of fragment size. Therefore, a sieving medium is required to separate DNA fragments. Use of a soluble medium is advantageous, because the buffer and the polymer can be treated in a similar fashion to a free solution separation and readily pumped into the capillary. Therefore, the injection of the sieving medium can be automated, and fresh sieving medium can be used with each CE run. Refilling capillaries also reduces the chance of contamination from previous samples that were analyzed in the same capillary. The sieving medium should have a viscosity that is sufficient to effectively separate the DNA fragments, but low enough to flow through the capillary. Additionally, the medium should not interact with other additives or the DNA.

Injection of DNA samples can be attained by two methods. Hydrodynamic injection entails the application of a pressure difference over the capillary. Alternatively, electrokinetic injection involves a combination of electrophoresis and EEO to inject the sample. The velocity of the DNA in the sample is dependent on the mobility and concentration of the other ions in the sample. While samples injected electrokinetically may be sensitive to other ions in the sample, band broadening is more apparent with hydrodynamic injections.

The preferential injection of higher charge-to-mass molecules (e.g., Cl^{-1} ions) with an electrokinetic injection affects, to a degree, the quantity of sample that can be injected into the capillary. This does not have a substantial effect on the ability to type STRs, because the samples can be diluted sufficiently.

Detection of DNA migrating past the detection window can be accomplished by monitoring UV absorption. However, the short pathway along the diameter of the capillary and dispersion of light by the capillary walls limit sensitivity of detection by UV absorbance. To increase sensitivity of detection, laser-induced fluorescence (LIF) is utilized. For STR typing, where single-stranded DNA molecules are separated by electrophoresis, fluorescent tags are covalently bound to the 5′ end of primers. Thus, a fluorescent molecule is incorporated into polymerase chain reaction (PCR) products and detected using LIF.

The AmpF*l*STR™ Profiler Plus™ (Profiler Plus) and AmpF*l*STR COfiler™ (Cofiler) DNA typing systems (both from PE Biosystems, Foster City, CA, USA) utilize PCR to amplify STRs to characterize DNA extracted from forensic specimens. The loci FGA, VWA, D3S1358, D5S818, D7S820, D8S1179, D13S317, D18S51, and D21S11 can be amplified by PCR simultaneously using the AmpF*l*STR Profiler Plus kit, and the loci CSF1PO,

TPOX, TH01, D3S1358, D7S820, and D16S539 can be amplified simultaneously using the AmpFlSTR COfiler kit. The loci D3S1358 and D7S820 are common to both kits. Appendix 2 lists the allele distributions for selected populations. Profiler Plus and Cofiler are multiplex systems that allow for the simultaneous amplification of a number of STR loci as well as a portion of the amelogenin gene located within the X and Y chromosomes. Analysis of amelogenin allows for gender determination. The Profiler Plus and Cofiler kits contain all reagents needed for amplification and include: primer sets that are specific for the various loci, PCR buffer and AmpliTaq Gold™ DNA polymerase (PE Biosystems), and the required allelic ladders. The primer sets contained within each kit consist of both unlabeled primers and those that are labeled with one of three distinctive fluorescent dyes. The use of multicolor dyes permits the analysis of loci with overlapping size ranges. The amplified fragments are separated according to size by capillary electrophoresis using the ABI PRISM® 310 Genetic Analyzer (PE Biosystems). The amplified fragments are detected by laser excitation, and the subsequent emission spectra are captured by a charge-coupled device (CCD) camera that displays the signals as peaks. The resulting data are graphically displayed as colored peaks noted by height in relative fluorescent units (RFUs) and time (scan number). This display is called an electropherogram.

The reference allelic ladders for each of the STR loci and reference fragments for amelogenin are also subjected to CE. These allelic ladders contain the more common alleles in the general population for each locus (Figure 1). Using the ladders, the alleles present in known and questioned DNA specimens may be determined.

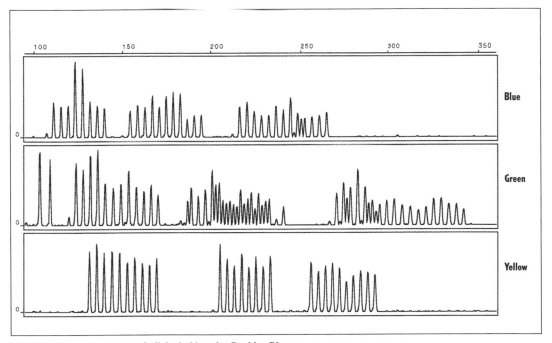

Figure 1. Electropherogram of allelic ladders for Profiler Plus system.

QUALITY CONTROL MEASURES

Quality assurance (QA) and quality control (QC) are important aspects of any system and/or protocol employed. Although this topic has yet to be discussed explicitly, it applies to all protocols and has been alluded to throughout the book. The QA protocols for DNA typing laboratories are described in Appendix 1. An example of QC considerations is displayed here to assist the scientist when implementing a procedure; QC is an extensive process. It is imperative that proper control samples be included when evidence samples are extracted, quantified, amplified, and typed through CE. The typing results obtained from these control samples are essential for the interpretation of STR and amelogenin typing results from evidentiary samples.

EXTRACTION CONTROLS

Reagent Blank

With each set of extractions of evidence samples a reagent blank is included. The reagent blank consists of all reagents used in the test process minus any sample and is processed through the entire extraction, quantification, amplification, and electrophoretic typing procedures. If more than one type of extraction procedure is used (with different extraction reagents), then a reagent blank is set up for each type of extraction or group of extraction reagents used. Twenty microliters of the reagent blank should be amplified.

AMPLIFICATION CONTROLS

Negative Control

A negative control is used with each set of amplifications. The negative control contains all reagents required for the amplification of DNA, except that no DNA is added. Twenty microliters of filter-purified water are placed in the negative control in lieu of a DNA solution. This control is processed through the amplification and capillary electrophoretic typing procedures.

Positive Amplification Controls

Cell line DNA 9947A is a positive control to evaluate the performance of amplification and CE. When the control specimen 9947A is amplified, the STR loci must exhibit the correct genotype. Additionally, specimen 9947A is a female control for amelogenin and should exhibit a single band at the 103 base peak.

A male control for amelogenin may be desirable and should exhibit 2 bands at positions corresponding with the 103 and 109 bases.

EQUIPMENT

The performance of the GeneAmp® PCR System 9700 Thermal Cycler (PE Biosystems) is critical for proper performance of the STR typing procedure; thus, the thermal cycler is evaluated on a regular basis.

THERMAL CYCLER—SYSTEM 9700

Periodic performance of the measures that are listed below will help ensure proper functioning of the thermal cyclers.

Cleaning

Refer to Cleaning the sample wells and Cleaning the sample block cover procedures described in the 96-Well Sample Block Module section of the GeneAmp PCR System 9700 User's Manual. Cleaning is performed quarterly.

Temperature Verification Test

Refer to Running the Temperature Verification Test procedures described in the 96-Well Sample Block Module section of the GeneAmp PCR System 9700 User's Manual Set. This test is performed quarterly.

Note that this test procedure requires the use of a Temperature Verification System (Part No. N801-0435; PE Biosystems) that is not furnished with the thermal cycler. This system is calibrated to NIST standards and is returned to PE Biosystems once per year for recalibration.

Temperature Non-uniformity Test

This procedure determines the temperature uniformity of the sample wells in the thermal cycler.

Refer to Running the Temperature Non-Uniformity Test procedures described in the Maintenance section of the User's Manual for the GeneAmp PCR System 9700.

Perform this test quarterly.

Rate Test and Cycle Test

These procedures verify the integrity of the cooling and heating system of a thermal cycler.

Refer to Running System Performance Diagnostics procedures described in the 96-Well Sample Block Module section of the GeneAmp PCR System 9700 User's Manual Set. These tests are performed quarterly.

PIPETS

Pipets should be exposed to UV light and washed with isopropanol as necessary. Maintenance and performance verifications are conducted annually, or as necessary.

ABI PRISM 310 GENETIC ANALYZER

Matrix

A new matrix is prepared, at a minimum, for each instrument semi-annually. Refer to How to Create the GeneScan Matrix File in the GeneScan® Analysis Software Experiments (PE Biosystems) of the ABI PRISM 310 Genetic Analyzer User's Manual for instructions on building a matrix. Once prepared, a matrix will be considered suitable for casework if a flat baseline is obtained when it is applied to the matrix standards.

Note: A new matrix must be built following cleaning, maintenance or replacement of a component of the optics.

ABI PRISM 310 Genetic Analyzer Response Evaluation

The performance of the ABI PRISM 310 Genetic Analyzer should be evaluated on a monthly basis with multiple injections of a standard control. The average peak height of a designated control allele is used as a standard. For example in the Federal Bureau of Investigation (FBI) Laboratory the D3S1358 "17" of a specified internal control sample must exhibit a peak height greater than 1000 RFU.

Capillary

A new capillary is installed after 100 injections.

Buffer and Waste Vials

The cathode buffer and waste vials are replaced, at a minimum, when the capillary is replaced. Each of these vials will be fitted with a new septum.

POWER MACINTOSH® COMPUTERS

Defragmenting Hard Drives

Hard drives are defragmented monthly using Norton Utilities.

QUALITY CONTROL OF CRITICAL SUPPLIES AND REAGENTS

Prior to use for case analysis, certain supplies and reagents should be checked against materials known to provide an acceptable result. The testing procedures for these items are described below.

AMPLIFICATION REAGENTS

Profiler Plus and Cofiler Kits

Each lot of kits is evaluated with template quantities of 9947A and other desired controls ranging from 60 pg to 1 ng DNA. The criteria for kit quality sufficient for use in casework are: *(i)* a minimum of 1 ng sensitivity, *(ii)* correct genotypes obtained for controls, and *(iii)* no artifactual peaks due to kit reagents. The 9947A sample will be suitable for use in casework if complete and correct typing of this sample is obtained with 1 ng template DNA. The allelic ladders are suitable for use in casework if: *(i)* each ladder allele is greater than 400 RFU, *(ii)* correct typing results are obtained for controls when the ladders are employed in the ABI PRISM GenoTyper® (PE Biosystems), and *(iii)* the ladder is devoid of additional peaks greater than 50 RFU.

CAPILLARY ELECTROPHORESIS REAGENTS

GS500[ROX]

Each lot of the internal standard GS500[ROX] is tested to ensure that: *(i)* the 75, 100, 139, 150, 160, 200, 300, 340, 350, and 400 base fragments are present and sized correctly, *(ii)* the control is correctly typed at the loci with this internal standard, *(iii)* the minimum peak height of fragments is 400 RFU, and *(iv)* no additional peaks greater than 50 RFU are present in the standard.

Formamide

Each lot of formamide is tested and must meet the criteria set forth for use in casework: *(i)* the conductivity of the formamide meets the manufacturer's specifications, *(ii)* the peak height of the control allele D3S1358 "17" must exhibit a peak height greater than a designated (e.g., 1000) RFU when injected with the lot of formamide, and *(iii)* correct typing of the control is obtained.

Following quality control testing, acceptable formamide is aliquoted and stored at -20°C. Aliquots can be stored at 4°C for a maximum of 2 weeks.

Performance Optimized Polymer 4™

Each lot of Performance Optimized Polymer 4 (POP-4) is considered of sufficient quality for use in casework if: *(i)* the GS500 400-base fragment elutes within 26 minutes, and *(ii)* correct typing of the control is obtained.

10× Genetic Analyzer Buffer with EDTA

Each lot of Genetic Analyzer Buffer with EDTA is considered acceptable for use in casework if: *(i)* the GS500 400-base fragment elutes within 26 minutes, and *(ii)* correct typing of the control is obtained.

9.1   DNA QUANTIFICATION PROCEDURE

Membrane

Each lot of nylon membrane is evaluated for use in casework.

Quantification Standards

Each lot of quantification standards is evaluated for use in casework.

Amplification

Materials and Reagents

- GeneAmp PCR System 9700 Thermal Cycler

- AmpF*l*STR COfiler and Profiler Plus PCR Amplification Kit.
- AmpliTaq Gold
- AmpF*l*STR PCR Mix
- Primer sets
- 9947A Cell Line DNA

Procedure

Note: All PCR setup steps must be performed in the laboratory where no amplified DNA is handled or stored, using hoods, reagents, and pipets dedicated to this area.

The reagents required for the amplification of the thirteen core STR loci are included in two kits, the AmpF*l*STR Profiler Plus PCR Amplification Kit and the AmpF*l*STR COfiler PCR Amplification Kit, or in a single combined kit, the AmpF*l*STR COfiler and Profiler Plus PCR Amplification Kit. For each amplification, combine DNA template, PCR mixture, primer set (Profiler Plus or Cofiler), and DNA polymerase (AmpliTaq Gold) in a reaction tube and subject this mixture to a series of controlled temperature changes in a GeneAmp PCR system 9700 (thermal cycler)

1. Turn on the thermal cycler. Program the thermal cycler according to the following instructions.

   a. Select option: Press F2 to select Create.

   b. Create a method: Press the down arrow to highlight 94.0 and change to 95.0 by pressing 9 5 0.

   c. Press the down arrow to highlight 5:00 and change to 11:00 by pressing 1 0 0.

   d. Continue through method using the arrows to change the appropriate times and temperatures to the following:

   > Hold at 95°C for 11 min
   > Cycle at 94°C for 1 min
   > at 59°C for 1 min
   > at 72°C for 1 min for 28 cycles
   > Hold at 60°C for 45 min
   > Hold at 25°C as desired

   e. Store method on instrument: Press F2 to select Store.

   This screen allows the selection of a user and to name a method.

   > Press F2 to select a user: Highlight correct user and press F1 to Accept. If a new user is desired, press F2 to select New. Use arrows and Enter key to name a user. Press F1 to Accept. Create a pin number, if so desired

   > Press F3 to select method: Press CE key to clear the method default. Use arrows and Enter key to name the method.

   > Press F1 to select Accept.

2. Determine the number of samples and controls to be amplified.

3. If samples have been stored at 4°C or frozen, allow the samples to thaw, then vortex-mix and spin the samples briefly in a microcentrifuge.

4. Transfer the DNA amplification reagents to the designated PCR setup area. Place the amplification reagent tubes in a dedicated rack that is not used for DNA preparation or amplified DNA handling.

5. Label the appropriate number of PCR tubes, including controls, and place the tubes in the amplification tray.

6. Prepare the amplification master mixture.

   a. Vortex-mix the AmpF*l*STR PCR mixture, AmpF*l*STR (Profiler Plus or Cofiler) primer set and AmpliTaq Gold DNA polymerase. Spin the tubes briefly in a microcentrifuge.

   b. The following volumes are per sample and include excess to allow for volume lost in pipetting. Multiply the volumes by the number of samples to be amplified including controls.

AmpF*l*STR PCR mixture	21.0 μL
Primer set	11.0 μL
AmpliTaq Gold	1.0 μL

c. Mix thoroughly and spin briefly in a microcentrifuge.

d. Aliquot 30 μL of master mixture per tube.

7. Each amplification is performed in a final volume of 50 μL. Use a new sterile pipet tip for the addition of each reagent or template DNA.

   a. Test Samples: Add 20 μL of template DNA (quantities to 2 ng) to each labeled tube containing master mixture. If a dilution of the template DNA is necessary, add a volume of sterile, filter-purified water to each tube that is equal to the difference between 20 μL and the volume of the template DNA.

   > For example, if 5 μL of the template DNA will be added, then: 20 μL - 5 μL = 15 μL water must be added.

   b. Reagent Blank Controls: Add 20 μL of the reagent blank to the designated master mixture tube(s).

   c. Positive Controls: Add 20 μL of 0.1 ng/μL human cell line (9947A) DNA provided in each kit to the designated master mixture tube.

   d. Negative Control: Add 20 μL of sterile, filter-purified water to the designated master mixture tube.

8. Cap the tubes tightly and push them down completely into the heat block of the thermal cycler fitted with a MicroAmp Tray. Maintain a record of the heat block position of each tube.

9. Begin the amplification process.

   a. Turn on the thermal cycler.

   b. Press F1 to Run.

   c. Highlight method.

   d. Press F1 to Start. Reaction volume is 50 μL, and ramp speed is 9600. Method will appear on screen.

10. After amplification, samples are stored at 4°C for up to 7 days or at -20°C for extended periods. Do not store amplified DNA in the same refrigerator as unamplified DNA samples or with DNA amplification reagents.

9.2   STR TYPING BY CAPILLARY GEL ELECTROPHORESIS

Materials and Reagents

- Power Macintosh computer
- ABI PRISM 310 Genetic Analyzer

- POP-4
- Capillary
- 10× Genetic Analyzer Buffer with EDTA
- Allelic ladders
- GeneScan-500 (internal standard)

- MicroAmp tray
- 0.2 mL strip tubes
- 96-well septa strip
- Retainer clip

9.3 | SETTING UP THE INSTRUMENT

Restart the Power Macintosh computer. Launch the ABI 310 Collection software if it is not already open.

The ABI PRISM 310 Genetic Analyzer is shown in Figure 2. The parts mentioned in the following sections are labeled.

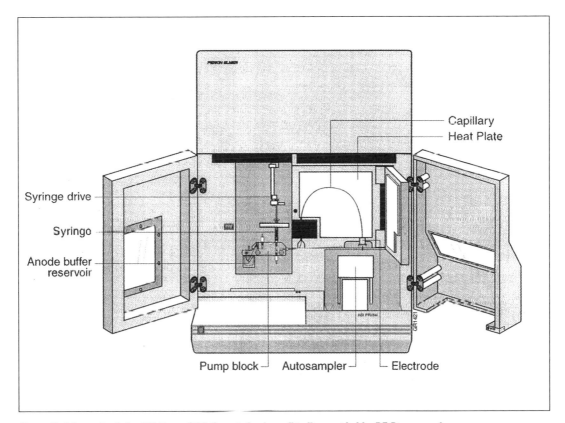

Figure 2. Schematic of the ABI PRISM 310 Genetic Analyzer (kindly provided by PE Biosystems).

Homing the Instrument

Procedure

1. Under the Window menu, select Manual Control.
2. Select Syringe Home from the Function pop-up menu, click execute.

3. Select Autosampler Home X, Y-Axis from the Function pop-up menu, click execute.

4. Select Autosampler Home Z-Axis from the Function pop-up menu, click execute.

Installing a New Electrode

Procedure

1. Unscrew the electrode thumbscrew located to the right of the detector door and just below the heat plate. Remove old electrode.

2. Insert the long end of the new platinum electrode into the electrode insertion hole of the thumbscrew.

3. Insert the hooked end of the electrode into the outer hole.

4. Replace the electrode back onto the instrument.

5. Under the Manual Control window, select Autosampler Home Z-Axis from the Function pop-up menu, click execute.

6. Use flush-cutting wire cutters (with the flat cutting surface facing the top of the instrument) to cut off a small piece of the electrode so that it is flush with the lower surface of the stripper plate (see Figure 3).

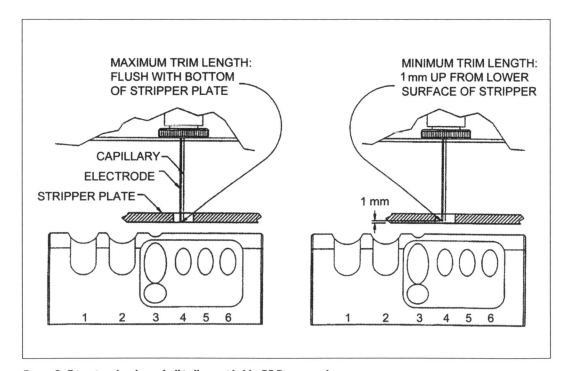

Figure 3. Trimming the electrode (kindly provided by PE Biosystems).

9.4 SETTING UP A RUN

Cleaning, Priming, and Loading the Syringe

Procedure

1. If the syringe drive is not in the home position, select Syringe Home from the Function pop-up menu of the Manual Control window.

2. Move the drive toggle to the left position to allow for removal of the syringe from the pump block.

3. Unscrew the syringe to remove it from the pump block.

4. Clean and rinse the 1.0-mL syringe with warm tap water (if necessary), then with filter-purified water. Remove all excess water from the inside and outside of the syringe using compressed air.

   Note: Place a small drop of distilled water on the tip of the plunger before inserting it back into the glass portion of the syringe. Without any water, the PTFE fitting of the plunger will cause excessive wear leading to leaks in the syringe.

5. Draw up approximately 0.1 mL of POP-4.

   Note: Before use, the POP-4 should be allowed to equilibrate to room temperature. If precipitate is present in the vial when removed from cold storage, it should go into solution at room temperature.

6. Pull the plunger to the 1.0-mL mark on the syringe. Invert the syringe several times to coat the walls with polymer.

7. Discard the polymer solution.

8. Fill the 1.0-mL syringe to a maximum of 0.8 mL of POP-4 polymer.

   Note: The following formula can be used as a guide for the approximate amount of polymer to sample:

   $$(Sample\ No. \times 6\ \mu L\ POP\text{-}4) + 200\ \mu L\ POP\text{-}4$$

   Note: Polymer that has been in the syringe on the instrument for more than 3 days should not be used for analysis. Discard such polymer. Do not return the unused polymer to the original bottle.

9. Rinse the outside of the syringe with distilled water and dry with a Kimwipe to remove excess polymer.

10. Remove any air bubbles by inverting the syringe and pushing out a small amount of polymer.

    Note: It is critical that all air bubbles be expelled from the syringe.

Cleaning the Pump Block

Procedure

1. Within the Manual Control window, select Syringe Home from the Function pop-up menu. Select Buffer Valve Open, press Execute.
2. Unscrew the glass syringe.
3. Remove any previously installed capillary.

   Note: If the capillary has been used for less than 100 injections, it may be stored with both ends in distilled water until future use.
4. Grasp the block with both hands and pull straight out.
5. Rinse the block, valves, and ferrule thoroughly with warm tap water (if necessary), then with filter-purified water.
6. Remove any excess water from inside and outside of the block using compressed air.
7. Slide the U-shaped end of the activator arm into the collar at the top of the plunger valve.

Installing the Syringe on the Pump Block

Procedure

1. Move the syringe drive toggle on the syringe pump drive to the left to allow for replacement of the syringe to the pump block.
2. Place the 1.0-mL syringe in through the right-hand port of the syringe guide plate and screw it into the pump block. The syringe should be finger-tight in the block.

   Note: Tighten the syringe at the metal collar. Excess pressure on the glass syringe barrel will break the seal of the collar, permanently damaging the syringe.
3. Hand-tighten the valves on the pump block to the left of and below the syringe.

Installation of a New Capillary

Procedure

1. Remove the new capillary from the plastic sleeve and clean the detection window with an isopropanol wipe.
2. Open the door covering the heat plate, then partially unscrew the plastic ferrule on the right side of the pump block.

3. Thread the capillary through the center of the ferrule.

4. Tighten the ferrule in the block. As the ferrule begins to set, adjust the end of the capillary so that it is positioned directly below the opening to the glass syringe. Finger-tighten the ferrule to secure the capillary.

5. Open the laser detector door and position the capillary in the vertical groove of the detector. Position the center portion of the capillary detection window in the laser detector window.

6. Close the laser detector door.

7. Thread the other end of the capillary through the capillary hole in the electrode thumbscrew until it extends approximately 0.5 mm beyond the electrode.

8. Secure the capillary in place by taping it to the heat plate above the laser detector door and above the electrode.

9. Close the heat plate door.

   Note: The capillary injection counter must be reset each time a new capillary is installed.

Calibrating the Autosampler

Procedure

   Note: Autosampler calibration should be performed whenever the capillary or electrode is changed or whenever the position of the electrode is altered.

1. Choose Autosampler Calibration from the Instrument menu.

2. Click start, follow the instructions on the screen to set the calibration point. When complete, click Done.

Fill the Buffer Reservoirs

Procedure

1. Dilute 1.5 mL of 10× Genetic Analyzer Buffer with EDTA to 15 mL (or a 1× solution) with filter-purified water.

   Note: A larger volume can be made if desired. The 1× solution can be stored at 4°C for up to 2 weeks.

2. Fill the anode buffer reservoir to the red line with 1× Genetic Analyzer Buffer with EDTA and replace the reservoir on the pump block.

3. Fill a 4-mL glass buffer vial to the full line with 1× Genetic Analyzer Buffer with EDTA. Insert the plastic vial lid with attached septum into the glass vial. Place this buffer vial in position 1 on the autosampler. This is the cathode buffer.

4. Fill a second 4-mL glass buffer vial to the fill line with filter-purified water. Insert the plastic vial lid with attached septum into the glass vial. Place this buffer vial in position 2 on the autosampler.

5. Fill a 1.5-mL microcentrifuge tube (cut off the lid) with filter-purified water and place into position 3 on the autosampler. Do not use a screw-top tube.

Note: To avoid electrical arcing, it is imperative that all surfaces of the buffer vials and the microcentrifuge tube be clean and dry.

Priming the Pump Block

Procedure

1. From the Manual Control window, select Buffer Valve Closed from the Function pop-up menu. Manually open the waste valve below the syringe.

2. Push down on the syringe plunger until a drop of polymer forms on the bottom of the waste valve. Tighten the waste valve to close.

3. Open the buffer valve by selecting Open Buffer Valve from the Function pop-up menu in the Manual Control Window.

4. Push down on the syringe plunger until all bubbles are removed from the polymer channel in the pump block. No air bubbles should remain.

5. Close the buffer valve by selecting Buffer Valve Close from the Function pop-up menu in the Manual Control window.

6. Move the syringe drive toggle to the right so that it is positioned over the syringe. Select Syringe Down from the Function pop-up menu in the Manual Control window. Select appropriate step intervals and click Execute to lower the syringe pump drive until the toggle is in contact with the plunger.

Set the Temperature

Procedure

1. Select Temperature Set from the Function pop-up menu in the Manual Control window.

2. Set the temperature to 60°C and click Execute. It takes approximately 30 min to reach the target temperature. Prepare the samples for injection during the heating step.

Note: Because a run will not be initiated until the heat plate has reached 60°C, performing this step at this time shortens the interval between run activation and execution.

Sample Preparation

Procedure

1. Prepare the appropriate number of sample tubes. Be sure to include tubes containing allelic ladder and GS500. Forty-eight samples or less could be set up in the 48-well sample tray using 0.5-mL Genetic Analyzer tubes. If more than 48 specimens are to be analyzed, setup in a MicroAmp tray retainer set using 0.2-mL strip tubes.

   Note: Both the 48-well and 96-well sample trays are keyed. Ensure that tube no. 1 is in position A1 in the sample tray and that tube no. 2 is in position A3 for the 48-well tray or position A2 for the 96-well tray, and so on.

2. Aliquot 24 µL of the formamide/GeneScan-500 [ROX] solution into the appropriate sample tubes.

3. Add 1.0 µL of PCR product or allelic ladder per tube.

4. Seal each tube with a septum.

   Note: Septa or Septa Strips melt at high temperatures. Do not autoclave or reuse septa.

   Note: Do not close the lid of the GeneAmp 9700 Thermal Cycler during the denaturing step. The septa will adhere to the lid.

   Note: In the 96-well sample tray, the retainer clip interferes with the insertion guides in the GeneAmp 9700 Thermal Cycler. Therefore, the retainer clip cannot be placed on the sample tray while using the 9700 to heat denature samples.

5. Using the 9700 thermal cycler, denature samples at 95°C for 3 min, then cool to 4°C for 3 min.

6. Secure the septa with a retainer clip.

Preparing for a Run

Procedure

1. Prepare a Sample Sheet. Choose New from the File menu.

2. Click on the appropriate GeneScan Sample Sheet icon (Figure 4).

3. Fill in the appropriate information as shown in Figure 5.

   Note: Ensure that the Std column highlights red.

   Note: Under the Sample Info column, ensure that each sample has a unique sample number, also ensure that each lane that contains an allelic ladder has the word "ladder" in the Sample Info column. This is necessary for automated allele calling with GenoTyper 2.0.

4. Select Save from the File menu.

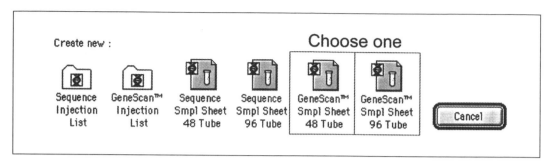

Figure 4. Icon menu.

#	Sample Name	Color	Std	Pres	Sample Info	Comments
A1	Profiler Plus Ladder	B		⊠	Profiler Plus Ladder	
		G		⊠	Profiler Plus Ladder	
		Y		⊠	Profiler Plus Ladder	
		R	◆	⊠	Profiler Plus Ladder	
A2	+ Control	B		⊠	+ Control	
		G		⊠	+ Control	
		Y		⊠	+ Control	
		R	◆	⊠	+ Control	
A3	– Control	B		⊠	– Control	
		G		⊠	– Control	
		Y		⊠	– Control	
		R	◆	⊠	– Control	
A4	700	B		⊠	700	
		G		⊠	700	
		Y		⊠	700	
		R	◆	⊠	700	
A5	701	B		⊠	701	
		G		⊠	701	
		Y		⊠	701	
		R	◆	⊠	701	

Sample Sheet "00021400015S" — GeneScan™ Sample Sheet

Figure 5. GeneScan Sample Sheet.

5. Select New from the File menu. Click on the GeneScan Injection List icon (Figure 6).

6. In the Injection List window that appears (see Figure 7), select the appropriate Sample Sheet from the Sample Sheet pop-up menu.

7. Ensure that the correct Module was inserted into the injection list. The correct parameters for Module GS STR POP4 (1 mL) F are:

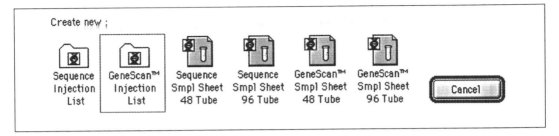

Figure 6. Icon menu.

Inj.	Tube & Sample Name	Module	Inj. Secs	Inj. kV	Run kV	Run °C	Run Time	Matrix file	Auto Anlz	Analysis Parameters
1	A1 – Profiler Plus Ladder	GS STR POP4	5	15.0	15.0	60	24	NED MATRIX 9-7	☐	
2	A2 – + Control	GS STR POP4	5	15.0	15.0	60	24	NED MATRIX 9-7	☐	
3	A3 – – Control	GS STR POP4	5	15.0	15.0	60	24	NED MATRIX 9-7	☐	
4	A4 – 700	GS STR POP4	5	15.0	15.0	60	24	NED MATRIX 9-7	☐	
5	A5 – 701	GS STR POP4	5	15.0	15.0	60	24	NED MATRIX 9-7	☐	
6	A6 – 702	GS STR POP4	5	15.0	15.0	60	24	NED MATRIX 9-7	☐	
7	A7 – 703	GS STR POP4	5	15.0	15.0	60	24	NED MATRIX 9-7	☐	
8	A8 – 704	GS STR POP4	5	15.0	15.0	60	24	NED MATRIX 9-7	☐	
9	A9 – 705	GS STR POP4	5	15.0	15.0	60	24	NED MATRIX 9-7	☐	
10	A10 – 706	GS STR POP4	5	15.0	15.0	60	24	NED MATRIX 9-7	☐	
11	A11 – 707	GS STR POP4	5	15.0	15.0	60	24	NED MATRIX 9-7	☐	
12	A12 – 708	GS STR POP4	5	15.0	15.0	60	24	NED MATRIX 9-7	☐	
13	B1 – 709	GS STR POP4	5	15.0	15.0	60	24	NED MATRIX 9-7	☐	
14	B2 – 710	GS STR POP4	5	15.0	15.0	60	24	NED MATRIX 9-7	☐	
15	B3 – 711	GS STR POP4	5	15.0	15.0	60	24	NED MATRIX 9-7	☐	
16	B4 – 712	GS STR POP4	5	15.0	15.0	60	24	NED MATRIX 9-7	☐	
17	B5 – 713	GS STR POP4	5	15.0	15.0	60	24	NED MATRIX 9-7	☐	

Sample Sheet: 0002140001SS   Run   Pause   Cancel
Length to Detector: 30 cm   Operator: AKJ

Figure 7. Injection list.

Inj. Secs:	5
Inj. kV:	15.0
Run kV:	15.0
Run °C:	60

Note: If it is necessary to reinject a sample with less PCR product, injection time may be decreased.

8. Set the run time to 26 min.

9. The first two injections are from the GS500 tube. Select injection one by clicking on the number one to the left of the first injection. Insert a row and copy the information from injection one to this row.

10. Place the samples on the Autosampler. Ensure that the instrument doors are closed.

Note: Ensure that the orientation of the sample rack is correct.

11. In the Injection List window, click the Run button to start the run.

9.5   GENESCAN ANALYSIS

Procedure

1. The sample files can be analyzed on any Power Macintosh computer containing GeneScan Analysis software version 2.1 or higher and Geno-Typer version 2.1 or higher. Open GeneScan Analysis.

2. Choose New from the File menu. Click the Project icon from the window that appears (Figure 8).

3. Select the appropriate Project file from the dialog box.

4. Within the Analysis Control Window, choose the blue, green, yellow, and

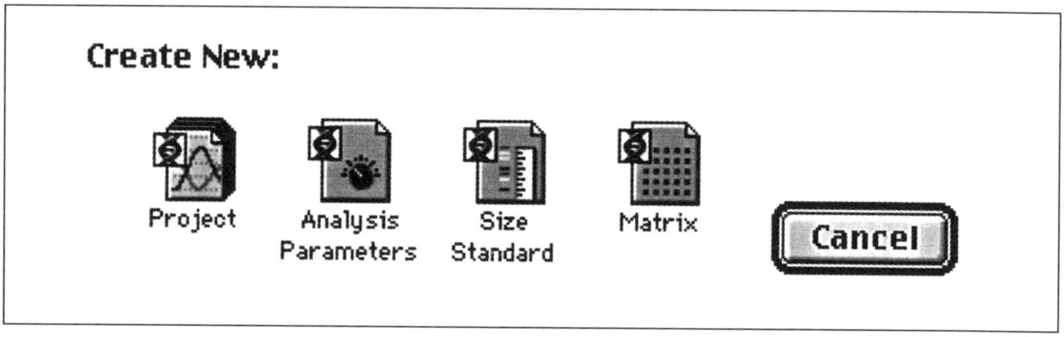

Figure 8. Icon menu.

red colors for analysis. Also, ensure that the red dye color is marked as the size standard (a diamond symbol should appear in the red boxes).

5. Select the Define New option from the Size Standard pop-up menu next to the sample you wish to use to define the new size standard. (For detailed information on using the Define New option see the GeneScan Analysis User's Manual). Define the standard peaks as in Figure 9.

6. Ensure that the analysis parameters are as in Figure 10.

   Note: It may be necessary to adjust the range of data points analyzed in the Analysis Range section of the Analysis Parameters window to capture all sizes from 75 to 400 bases.

7. Click Analyze.

8. View the results from the Results Control window.

9. For each sample, verify that all peaks in the GeneScan-500 size standard have been correctly assigned. If incorrect peak assignment is observed, define a new size standard using the GS500 from that lane.

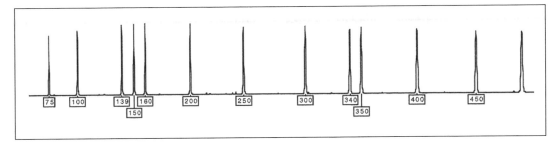

Figure 9. GS500.

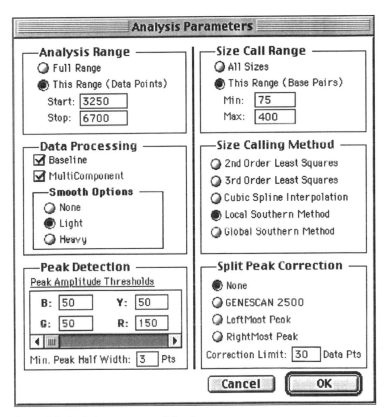

Figure 10. Analysis parameters dialog box.

9.6 | GENOTYPER ANALYSIS

Before Applying Genotyper

Procedure

1. The Sample Info column of the GeneScan sample sheet must contain a unique sample description (see Figure 11) in order for GenoTyper soft-

#	Sample Name	Color	Std	Pres	Sample Info	Comments
					GeneScan™ Sample Sheet	
A1	Profiler Plus Ladder	B		☒	Profiler Plus Ladder	
		G		☒	Profiler Plus Ladder	
		Y		☒	Profiler Plus Ladder	
		R	◆	☒	Profiler Plus Ladder	
A2	+ Control	B		☒	+ Control	
		G		☒	+ Control	
		Y		☒	+ Control	
		R	◆	☒	+ Control	
A3	– Control	B		☒	– Control	
		G		☒	– Control	
		Y		☒	– Control	
		R	◆	☒	– Control	
A4	700	B		☒	700	
		G		☒	700	
		Y		☒	700	
		R	◆	☒	700	
A5	701	B		☒	701	
		G		☒	701	
		Y		☒	701	
		R	◆	☒	701	

Sample Sheet "0002140001SS"

Figure 11. GeneScan sample sheet.

ware to build a table. Also, the lanes or injections that contain the allelic ladder must contain the word "ladder" in the Sample Info column of the sample sheet. The GenoTyper template identifies the allelic ladder by searching for the word "ladder" in Sample Info.

2. If the Sample Info field in the sample sheet was not completed before initiating a run, it is possible to add this information once the data is in the GenoTyper application.

a. Under the View menu, choose show Dye/lanes Window.

b. Select the first sample by clicking on the row.

c. Select the Sample Info box at the top of the window and type the sample description.

d. Repeat the above steps for every sample in the Dye/lanes list. Enter the same sample description for each of the dye colors for a single sample.

9.7   USING A TEMPLATE FILE

Procedure

1. Double click on the template file (either Profiler Plus or Cofiler) to simultaneously open the GenoTyper application and the template file.

   Note: The template file is a Stationary Pad, therefore, a new document is created each time it is opened. The original template file is not over-written.

2. Import GeneScan sample files by selecting Import GeneScan File(s) from the File menu.

3. Select the check boxes for Blue, Green, Yellow, and Red. This will import the analyzed data for all colors. Import Raw Data.

4. Select the project file and click import. Individual sample files may be imported by selecting the file and clicking import.

5. If each sample does not have the Sample Info completed, see steps 1 and 2 under Before Applying GenoTyper.

6. From the Macro list at the bottom left of the Main Window, select Check GS500.

7. Under the Macro menu, choose Run Macro. In the Plot window that appears, scroll through each sample and verify that each peak of GeneScan-500 was assigned the correct size by GeneScan Analysis software.

8. From the Macro list at the bottom left of the Main Window, select the Kazam 4%.

9. Under the Macro menu, choose Run Macro. The macro will take several minutes to complete. When finished, a plot window opens containing the labeled alleles for the blue data. Examine the data as described in the Examining Data section.

10. In the Main Window, click the G (Green data) button in the upper left corner. Under the views menu, select Show Plot Window. Examine the data as described in the Examining Data section.

11. In the Main Window, click the Y (Yellow data) button in the upper left corner. Under the views menu, select Show Plot Window. Examine the data as described in the Examining Data section.

12. Print the data.

9.8   EXAMINING DATA

Procedure

1. Ensure that the peaks of the allelic ladder are labeled with the correct allele designations. See Figures 12 and 13.

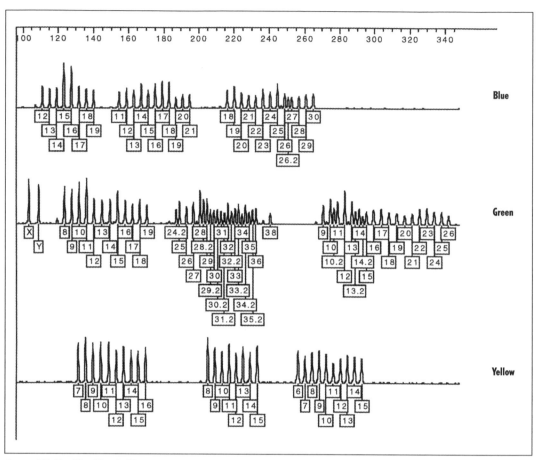

Figure 12. Profiler Plus ladder.

2. Peak labeling criteria are as follows:

    a. Allele categories, which appear as dark gray bars in the plot window, are defined to be ± 0.5 bases wide. Peaks that fit into the ± 0.5 bases window of an allele category will be labeled with that allele designation.

    Note: The categories for THO1 9.3 and 10 are ± 0.4 bases wide.

    b. Peaks that do not size within an allele category will have a label indicating OL Allele (off ladder allele).

    Note: The Profiler Plus and Cofiler templates contain allele bins for alleles represented in the allelic ladder as well as other known alleles.

    c. The Kazam 4% macros include a step that removes labels from peaks that are less than 4% of the largest allele present at each locus.

    d. Peaks that are less than the minimum peak height previously specified in GeneScan software (50 RFU for blue, green, and yellow; 150 RFU for red) cannot be labeled. These peaks will not line up by size appropriately in the GenoTyper plot window.

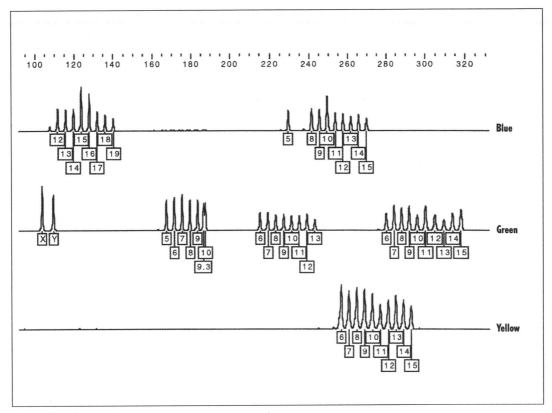

Figure 13. Cofiler ladder.

3. Scroll through the remaining samples below the allelic ladder (see Figure 14) to examine the peak labels and edit the peak labels where necessary. Clicking on a labeled peak will remove the label. Clicking on the same peak again will label the peak by size (bases).

Note: To change a label, choose Change Label from the Analysis menu and select the desired label (peak height, category's name, size, etc.).

4. Print edited data.

Item	Name	Description
1	Dye/lane list	Shows specific dye/lanes available for analysis.
2	Upper Graphical Area	Shows electropherogram plots.
3	Lower Graphical Area	Shows peak labels.
4	Category list	Shows criteria for a group of peaks selected on the basis of parameters you define using the category features.

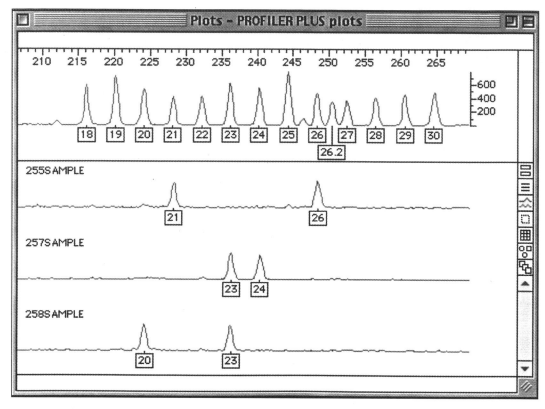

Figure 14. GenoTyper Plot window.

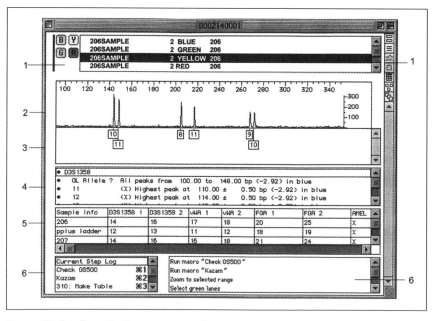

Figure 15. GenoTyper Application main window.

5	Table Area	Shows tabular data for created tables.
6	Macro List	Lists the names of macros that you have created and can run.
7	Window selection buttons	Open windows for a particular GenoTyper document, and provides access to other application software.
8	Step list	Contains the list of steps for the current Step Log or the macro selected in the Macro list.

9.9 ALLELE DESIGNATION

Designation of Profiler Plus and Cofiler Alleles

The STR and amelogenin fragments are separated according to size by CE. During this process, each fragment travels 36 cm to an argon laser that excites the dyes at 485 and 515 nm. A CCD camera captures the subsequent emission spectra of these dyes, and the signal is displayed as peaks. The fluorescent tag attached to one of the primers determines the color of the peak. The peak height is a measure of the amount of fluorescence detected. The electropherogram is a graphical display of colored peaks measured by height in RFUs and time (scan number).

The size of each allele is determined in GeneScan Analysis by comparison to the internal standard GS500 using Local Southern analysis. GS500 is labeled with ROX and is displayed as red peaks. The GenoTyper Software is used to generate allelic designations from sized GeneScan data. This is accomplished by comparison to the allelic ladders. The GenoTyper software recognizes the first allele of the ladder and creates 1 base bins around each allele in the ladder. The bins for the TH01 9.3 and 10 alleles are 0.8 bases wide. Sample peaks are labeled by comparison of their size to the size of the ladder bins. The ladders generally are composed of the more common alleles in the general population. The ladder does not carry all alleles. There are a few virtual bins created in GenoTyper. If an allele falls into a virtual bin, an allelic designation will be assigned. However, if an allele is not represented by either a bin created by the ladder or a virtual bin, the GenoTyper software will designate this peak as OL for off ladder allele. Off ladder alleles will be designated in accordance with guidelines of the International Society for Forensic Haemogenetics.

Designation of the Amelogenin Fragments

Amelogenin typing is performed by comparing fragments to the ladder. The X allele is approximately 103 bases, the Y allele is approximately 109 bases. A specimen of male origin will exhibit both the X and Y amelogenin fragments, whereas a specimen of female origin will exhibit only the X fragment.

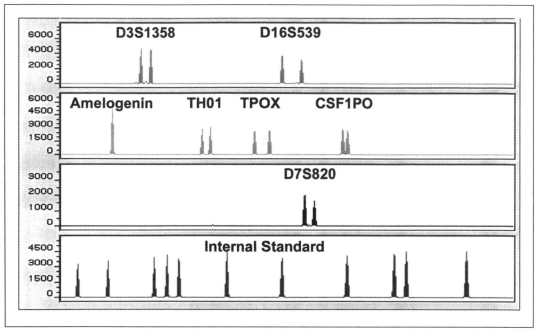

Figure 16. The Cofiler system enables simultaneous amplification of six STR loci and the amelogenin locus. This sample was analyzed on ABI Prism 310 Genetic Analyzer.

Preliminary Evaluation of Data

Procedure

As described above, allelic designations are computer assisted. To ensure accuracy of genotyping, the following criteria must be met.

1. The internal lane marker GS500 must have correct sizes assigned to the peaks used for sizing. It is noted that the 250 base peak is not used for sizing purposes. The 75 and 350 base peaks must be captured for all samples. The 400 base peak must be captured for the D18S51 26 allele or other alleles greater than 340 bases in size.

2. All ladder alleles must be correctly called in GenoTyper and be free of spikes.

3. Peak heights must be 200 RFU or greater to be considered conclusive for match purposes. If peak heights fall below 200 RFU but offer exculpatory information, or indicate the presence of a mixture, then this information will be reported.

4. Spikes (i.e., anomalous and nonreproducible signals) are potential artifacts of the CE process. Generally, spikes are thin peaks with large peak heights, are present in two or more colors, and have the same scan number in each color. If a sample has a spike at the 75 base peak, between 100 and 350 base peaks, or at 400 base peak, the sample may be rein-

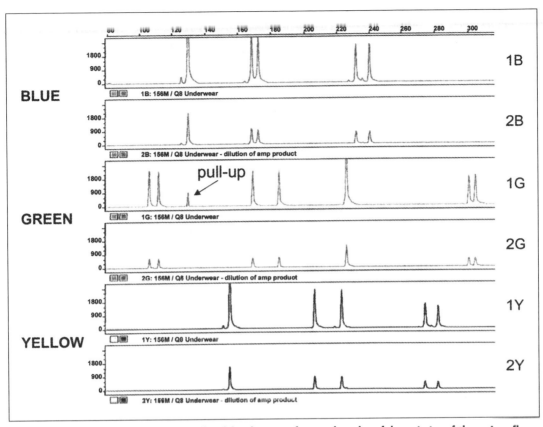

Figure 17. Peaks that are too high (i.e., off-scale) or because of spectral overlap of the emission of the various fluors may cause pull-up. 1B (Blue), 1G (Green), and 1Y (Yellow) are the three fluors used to label the multiplex loci of the one sample. The large blue peak furthest to the left of 1B is causing pull-up of a peak in the green (1G and designated by the arrow). In this example, when the same sample is diluted (2B, 2G, and 2Y), pull-up is eliminated.

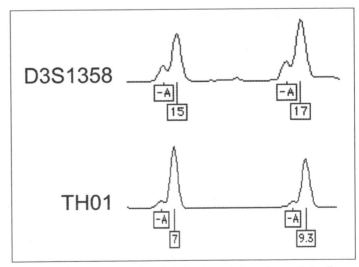

Figure 18. Chromatogram of two loci showing positions of true alleles (large peaks) and -A products (small shoulder peaks).

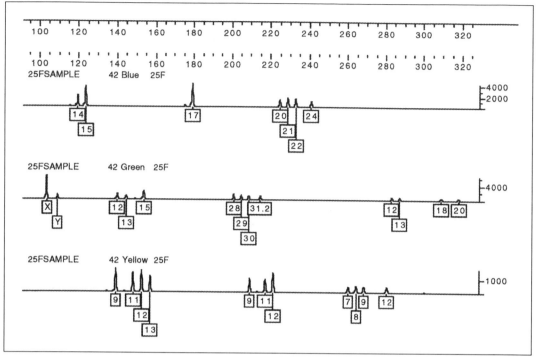

Figure 19. Profiler Plus results from sexual assault evidence. There are two contributors in this sample, which is evident by more than two peaks at some of the loci and the notable imbalance of the peak heights at some loci.

jected. If a spike is visible within GeneScan Analysis, but not in Geno-Typer, the GeneScan electropherogram should be printed.

5. Pull-up Peaks. Excess input DNA may cause the appearance of pull-up peaks because the fluorescent intensity exceeds the linear dynamic range of the instrument. The data will be off scale and may result in raised baselines or pull-up of one or more colors under the off scale peaks. Samples exhibiting pull-up may be reinjected with less DNA.

6. Bleed-through. Cross talk between data in adjacent channels (e.g., green and blue) may result in bleed-through. The effect of bleed-through peaks is similar to pull-up, however all data are on scale. Samples exhibiting bleed-through may be reinjected.

7. Samples with off-scale data (8191 RFU seen in unanalyzed data) may be diluted and reinjected if necessary.

8. Within a single source sample, genotypes generated for the loci D3S1358 and D7S820 in the Profiler Plus and Cofiler amplification and typing systems must be concordant.

9. Off Ladder alleles that fall between alleles within the ladder will be designated in accordance with guidelines of the International Society for Forensic Haemogenetics. Off ladder (OL) calls are first converted to size in bases, then compared to the size of the appropriate ladder alleles, and the allelic designation is finally determined. If the OL is not a perfect

repeat, but rather varies by 1, 2, or 3 bases from a ladder allele, then it will be designated as an integer of that variation. For example, if a green OL peak size is 238.39 bases, and the 36 allele of the D21S11 ladder is 236.32 bases, then the peak will be designated a D21S11 36.2. If an allele falls above the largest or below the smallest peak of the sizing ladder, the allele will be designated as either greater than (>) or less than (<) the respective ladder allele.

SUGGESTED READING

1. Anker, R., T. Steinbrueck, and H. Donis-Keller. 1992. Tetranucleotide repeat polymorphism at the human thyroid peroxidase (hTPO) locus. Hum. Mol. Genet. *1*:137.

2. Budowle, B., T. Moretti, K. Keys, B. Koons, and J. Smerick. 1997. Validation studies of the CTT multiplex system. J. Forensic Sci. *42*:701-707.

3. Buel, E., M. Schwartz, and M. LaFoutain. 1998. Capillary electrophoresis STR analysis: comparison to gel-based systems. J. Forensic Sci. *43*:164-170.

4. Clark, J. 1988. Novel non-templated nucleotide addition reactions catalyzed by procaryotic and eukaryotic DNA polymerases. Nucleic Acids Res. *16*:9677-9686.

5. Crouse, C.A. and J. Schumm. 1995. Investigation of species specificity using nine PCR-based human STR systems. J. Forensic Sci. *40*:952-956.

6. Edwards, A., A. Civitello, H. Hammond, and C.T. Caskey. 1991. DNA typing and genetic mapping with trimeric and tetrameric tandem repeats. Am. J. Hum. Genet. *49*:746-756.

7. FBI Laboratory. 1997. Guidelines for Operation of the Combined DNA Index (CODIS). Washington, D.C.

8. Fowler, J.C.S., L.A. Burgoyne, A.C. Scott, and H.W.J. Harding. 1988. Repetitive deoxyribonucleic acid (DNA) and human genome variation—a concise review relevant to forensic biology. J. Forensic Sci. *33*:1111-1126.

9. Fregeau, C.J. and R.M. Fourney. 1993. A DNA typing with fluorescently tagged short tandem repeats: a sensitive and accurate approach to human identification. BioTechniques *15*:100-119.

10. Guttman, A. and N. Cooke. 1991. Effect of temperature on the separation of DNA restriction fragments in capillary gel electrophoresis. J. Chromatogr. *559*:285-294.

11. Isenberg, A.R., B.R. McCord, B.W. Koons, B. Budowle, and R.O. Allen. 1996. DNA typing of a polymerase chain reaction amplified D1S80/amelogenin multiplex using capillary electrophoresis and a mixed entangle polymer matrix. Electrophoresis *17*:1505-1511.

12. Issaq, H., K. Chan, and G. Muschik. 1997. The effect of column length, applied voltage, gel type, and concentration on the capillary electrophoresis separation of DNA fragments and polymerase chain reaction products. Electrophoresis *18*:1153-1158.

13. Kimpton, C., P. Gill, A. Walton, A. Urquhart, E. Millican, and M. Adams. 1993. Automated DNA profiling employing multiplex amplification of short tandem repeat loci. PCR Methods Appl. *3*:13-22.

14. Kimpton, C.P., A. Walton, and P. Gill. 1992. A further tetranucleotide repeat polymorphism in the vWF gene. Hum. Mol. Genet. *1*:287.

15. Kline, M., D. Duewer, P. Newall, J. Redman, D. Reeder, and M. Richard. 1997. Interlaboratory evaluation of short tandem repeat triplex CTT. J. Forensic Sci. *42*:897-906.

16. Lazaruk, K., P. Walsh, F. Oaks, D. Gilbert, B. Rosenblum, S. Menchen, D. Scheibler, H. Wenz, C. Holt, and J. Wallin. 1998. Genotyping of forensic short tandem repeat (STR) systems based on sizing precision in a capillary electrophoresis instrument. Electrophoresis *19*:86-93.

17. Li, H., L. Schmidt, M.H. Wei, T. Hustad, M.I. Lerman, B. Zbar, and K. Tory. 1993. Three tetranucleotide polymorphisms for loci: D3S1352, D3S1358, D3S1359. Hum. Mol. Genet. *2*:1327.

18. Lincoln, P. 1997. DNA recommendations—further report of the DNA Commission of the ISFH regarding the use of short tandem repeat systems. Forensic Sci. Int. *87*:179-184.

19. Mayrand, P.E., K.P. Corcoran, J.S. Ziegle, J.M. Robertson, L.B. Hoff, and M.N. Kronick. 1992. The use of fluorescence detection and internal lane standards to size PCR products automatically. Appl. Theor. Electrophor. *3*:1-11.

20. Mayrand, P.E., J. Robertson, J. Ziegle, L.B. Hoff, L.J. McBride, J.S. Chamberlain, and M.N. Kronick. 1991. Automated genetic analysis. Ann. Biol. Clin. *4*:224-230.

21. Mills, K.A., D. Even, and J.C. Murray. 1992. Tetranucleotide repeat polymorphism at the

human alpha fibrinogen locus (FGA). Hum. Mol. Genet. *1*:779.

22. **Moller, A., E. Meyer, and B. Brinkmann.** 1994. Different types of structural variation in STRs: Hum FES/FPS, HumVWA and HumD21S11. Int. J. Legal Med. *106*:319-323.

23. **PE Biosystems.** 1995. ABI PRISM® 310 Genetic Analyzer User's Manual. Rev. 1. Foster City, CA.

24. **PE Biosystems.** 1996. ABI PRISM® GeneScan® Analysis Software 2.1 User's Manual. Foster City, CA.

25. **PE Biosystems.** 1996. ABI PRISM® GenoTyper® 2.0 User's Manual. Foster City, CA.

26. **PE Biosystems.** 1997. GeneAmp® PCR System 9700 User's Manual Set. Foster City, CA.

27. **PE Biosystems.** 1998. ABI PRISM® 310 Genetic Analyzer User's Manual. Foster City, CA.

28. **PE Biosystems.** 1998. AmpFlSTR™ COfiler™ PCR Amplification Kit User Bulletin. Foster City, CA.

29. **PE Biosystems.** 1998. AmpFlSTR™ Profiler Plus™ PCR Amplification Kit User's Manual. Foster City, CA.

30. **Polymeropoulos, M.H., D.S. Rath, H. Xiao, and C.R. Merrill.** 1991. Tetranucleotide repeat polymorphism at the human tyrosine hydroxylase gene (TH). Nucleic Acids Res. *19*:3753.

31. **Puers, C., H. Hammond, L. Jin, C. Caskey, and J. Schumm.** 1993. Identification of repeat sequence heterogeneity at the polymorphic short tandem repeat locus HUMTH01 [AATG]n and reassignment of alleles in population analysis by using a locus-specific allelic ladder. Am. J. Hum. Genet. *53*:953-958.

32. **Sharma, V. and M. Litt.** 1992. Tetranucleotide repeat polymorphism at the D21S11 locus. Hum. Mol. Genet. *1*:67.

33. **Siles, B., G.B. Collier, D.J. Reeder, and W.E. May.** 1996. The use of a new gel matrix for the separation of DNA fragments: a comparison study between slab gel electrophoresis and capillary electrophoresis. Appl. Theor. Electrophor. *6*:15-22.

PCR-Based Analysis: Mitochondrial DNA Sequencing

10

MITOCHONDRIAL DNA

Most of the DNA in human cells is contained within the 46 chromosomes in the nucleus. Outside the nucleus in the cytoplasm are mitochondria, which are subcellular organelles that contain an extrachromosomal genome separate and distinct from the nuclear genome. Human mitochondrial DNA (mtDNA) differs from nuclear DNA as follows: *(i)* It exists as a closed circular, rather than linear, genome. *(ii)* The mtDNA genome is smaller, consisting of approximately 16.5 kb. *(iii)* It is more compact, containing coding sequences only for 2 ribosomal RNAs, 22 transfer RNAs, 13 proteins, and a noncoding region approximately 1100 bp long, called the displacement loop (D-loop) or control region (Figure 1). *(iv)* All mtDNA is inherited maternally. *(v)* mtDNA does not undergo recombination. *(vi)* It is present in high copy number in a cell.

The mtDNA genome has been completely sequenced. The double-stranded mtDNA molecule is comprised of one purine-rich strand and one pyrimidine-rich strand, designated the heavy (H) chain and light (L) chain, respectively. Nucleotide positions in the mtDNA genome are numbered according to the convention of Anderson et al. (1981) (1) (See Appendix 2). An arbitrary position on the heavy strand begins the numerical designation of each base pair, continuing around the circle for approximately 16 569 bp.

Unlike nuclear DNA, mtDNA is maternally inherited. Barring mutation, the mtDNA sequence is identical for siblings and all their maternal relatives (Figure 2). This characteristic can be helpful in forensic cases, such as analysis of the remains of a missing person, where known maternal relatives can provide reference samples for direct comparison to the questioned mtDNA type.

DNA Typing Protocols: Molecular Biology and Forensic Analysis
By B. Budowle, J. Smith, T. Moretti, J. DiZinno
©2000 Eaton Publishing, Natick, MA

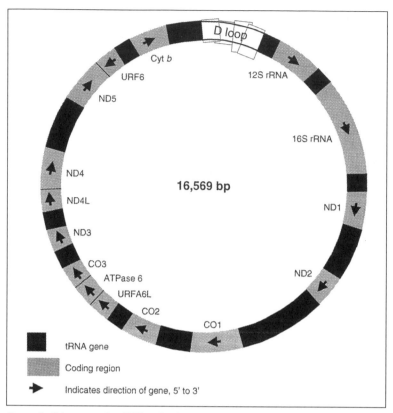

Figure 1. Schematic of mtDNA molecule.

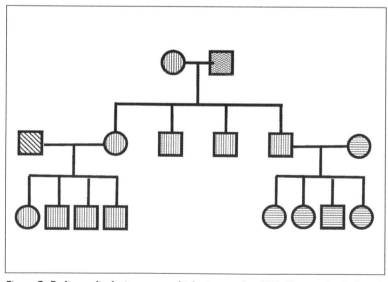

Figure 2. Pedigree displaying maternal inheritance of mtDNA. Those individuals with the same fill-in pattern carry the same mtDNA sequence. Note: Barring mutation, all maternal relatives have the same mtDNA sequence.

An important feature of mtDNA, which simplifies DNA sequencing, is its monoclonal nature. However, a condition, known as heteroplasmy, is encountered. A person is considered heteroplasmic if he/she carries more than one mtDNA type. Careful analysis and direct comparisons between multiple known samples and a questioned sample should, in most cases, alleviate interpretational difficulties that may arise due to the presence of heteroplasmy.

The main advantage of typing mtDNA, as opposed to nuclear DNA, for forensic identification is the high copy number of mtDNA molecules in a cell. In cases in which the amount of extracted DNA is very small or degraded, as in tissues such as bone, teeth, and hair, the probability of achieving a DNA typing result from mtDNA is higher than that of polymorphic markers found in nuclear DNA.

The mtDNA does not undergo recombination. However, the low fidelity of mtDNA polymerase and the apparent lack of mtDNA repair mechanisms have led to a higher rate of mutation in the mitochondrial genome compared with the nuclear genome. Some regions of the mtDNA genome appear to be evolving at 5 to 10 times the rate of single copy nuclear genes. It is these regions that are of interest for human identity testing. It has been estimated that among unrelated Caucasians, there is an average of eight nucleotide differences between individuals. This average is higher in individuals of African descent, being approximately 15 bases differing between individuals.

An important potential application of mtDNA sequencing in forensics is the analysis of hair shafts. Since individual hairs contain very small quantities of DNA, mtDNA sequence analysis may be the only viable technique for analysis. In fact, sequences can be obtained from as little as 1 to 2 cm of a single hair shaft. Sequencing of mtDNA from hair shaft, bones, and teeth is performed routinely at the FBI to assist in resolving violent crime cases.

DNA SEQUENCING

Sequencing entails determining the array of nucleotides along the DNA molecule. The reagents required to perform sequencing (i.e., the Sanger method) are: purified, single-stranded DNA template (which can be obtained from an amplified sample), an oligonucleotide primer, four deoxyribonucleoside triphosphates (dATP, dCTP, dGTP, and dTTP—collectively known as dNTPs), four dideoxyribonucleoside triphosphate analogs (ddATP, ddCTP, ddGTP, and ddTTP), DNA polymerase, and buffer. Polymerase chain reaction (PCR) is used to generate the DNA template for sequencing. Following denaturation of the amplified template, a sequencing primer is annealed to the single-stranded DNA. The primer is extended across the target sequence by the sequential addition of the four dNTPs and one terminator ddNTP, as catalyzed by a DNA polymerase. The ddNTP in the sequencing reaction is incorporated into the growing chain by complementary base pairing to the template, as are dNTPs, and therefore competes with its dNTP analog for incorporation. However, chain elongation is terminated at the point where the ddNTP is incorporated since, unlike dNTPs, ddNTPs do not have a 3′ hydroxyl group that is necessary for chain elongation (Figure 3).

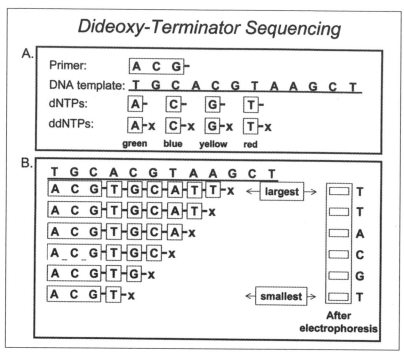

Figure 3. Schematic displaying the basic process of cycle sequencing.

Traditionally, the Sanger sequencing reaction is carried out in four separate tubes, with each tube containing a different ddNTP. Thus, the fragments in each tube terminate with a particular base that corresponds to the ddNTP. Specifically, for example, one of the four tubes contains dATP, dCTP, dTTP, dGTP, and ddATP. The sequencing chain is extended, with either dATP or ddATP added complementary to a thymine in the template. If a dATP is incorporated, the chain can continue to be elongated. In contrast, if a ddATP is incorporated, the chain is terminated. The result is a collection of fragments in the tube that differ in length, but all terminate with ddATP. Similarly, ddCTP is used, instead of ddATP, in another tube, resulting in a collection of differently sized fragments that terminate with ddCTP. The third and fourth tubes contain fragments that terminate in ddGTP and ddTTP, respectively. The reaction mixtures from the four tubes are placed in separate, adjacent lanes of a denaturing polyacrylamide gel, and the fragments are separated by electrophoresis. The result is a ladder of fragments, across the four lanes, that differ in length by one nucleotide. The fragments in each lane terminate in the ddNTP that corresponds to that which was added in the reaction mixture. Therefore, there is an A lane, a C lane, a G lane, and a T lane. The sequence is read up the ladder in the four lanes.

Typically, one of the dNTPs in the reaction mixture is radiolabeled, and hence all sequencing fragments are internally labeled prior to electrophoretic separation. Following electrophoresis, the sequencing gel is exposed to film. After autoradiography, the fragments can be visualized, and the DNA sequence of the target DNA can be determined. This approach to sequencing is known as manual sequencing; while effective, the process is time-consuming and labor-intensive, particularly for data interpretation.

Instead of manual sequencing, PCR-amplified DNA templates can be sequenced with the use of a fluorescent detection-based automated sequencer. These computer-driven instruments provide high-throughput and ease of data management. As with manual sequencing, fluorescent-based systems also rely on ddNTP incorporation. However, with fluorescence detection, the extended fragments are labeled by incorporation of either a fluorescently labeled oligonucleotide primer or a ddNTP with a fluorescent dye attached. Since each primer or each ddNTP can be labeled with different fluorescent dyes, all four termination reactions can be performed together (in the case of labeled ddNTPs) or performed separately, pooled (in the case of labeled primers) prior to electrophoresis, and loaded in a single lane of a sequencing gel. The temporal, sequential passage of the fluorescently labeled fragments past a detector window enables determination of the sequence of the template DNA. The fluorescence approach is attractive because of the ease of manipulation and analysis, reduction in labor, and higher sample throughput.

POST-AMPLIFICATION QUANTITATION

To improve the quality of mtDNA sequencing, the appropriate quantity of amplified template molecules for the cycle sequencing reaction should be determined. Capillary electrophoresis (CE) can be used to quantitate the amplicons; thus, the quality of results in mtDNA sequencing is augmented. This strategy also allows for the quality of the purified PCR products to be evaluated to determine if nonspecific PCR products were produced or if PCR primers were not completely removed during purification. The CE strategy exploits two parameters: the use of an internal standard and the use of an intercalating dye to detect double-stranded DNA. The PCR products are diluted with deionized water, which contains a reference standard, and are injected hydrodynamically. Detection occurred by means of laser-induced fluorescence of the DNA-intercalating dye complex. The ratio of the PCR product peak area to the reference standard peak area enables determination of the quantity of PCR products in the PCR.

Another benefit of the capability of quantifying the amplified mtDNA product is to better address potential contamination. The sensitivity of detection of mtDNA typing is greater than other PCR-based assays described in Chapters 6 through 9. Thus, low level contamination that typically is of little concern for most other PCR-based assays must be considered when typing mtDNA. Low level contamination, at times, will be observed in the negative controls during mtDNA analysis. These low levels often have no impact on the reliability of the results obtained. However, the level of contamination must be monitored so it does not exceed acceptable levels. Because of the sensitivity of mtDNA typing and the incorporation capacity of dideoxynucleotides by the DNA polymerase, it has been determined that contamination in the negative controls is less than 10% of the quantity of DNA observed in the evidence samples and will not affect the sequencing results.

QUALITY ASSURANCE/QUALITY CONTROL

The same general guidelines described in previous chapters (more so in Chapter 9, entitled PCR-Based Analysis: Capillary Electrophoresis for Short Tandem Repeat Loci Typing and Real Time Fluorescent Detection) apply to mtDNA typing, and reduction of potential contamination within the laboratory is extremely important for mtDNA analyses. The practices include: use of laboratory coats, gloves, and aerosol resistant tips—only one case and only one item from that case should be open at a time. Items of evidence (when possible) should be extracted and amplified prior to known reference samples, pre- and post-amplification areas should be physically separated, work surface areas should be thoroughly cleaned with 10% bleach after each case, work spaces under dedicated hoods should be exposed to UV light for a minimum of 15 minutes, all appropriate reagents should be exposed to UV light for at least 15 minutes, UV sterilization in hoods and Stratalinker® UV Crosslinkers (Stratagene, La Jolla, CA, USA) should be monitored monthly using UVICIDE® cards (Vanguard International, Neptune, NJ, USA), etc.

In addition to autoclaving and/or exposing to UV light all possible glassware, plasticware, and reagents, dedicated hoods, laboratory coats, and pipets are used. Disposable gloves are a requisite. All bone and teeth extractions should be performed under a ventless air filtration hood with a HEPA filter [Pre Filter # FE 4013 for hood FE 2600 (Misonix, Farmingdale, NY, USA)]. Filter should be changed every three months.

Reagent blanks and negative controls are used to monitor contamination levels. If the background contamination level exceeds 10% of the quantity of amplified mtDNA in the evidence sample, the extraction will be repeated, and a new control will be prepared. Regardless of the relative quantity of amplified DNA in the reagent blank or negative control, if a typeable sequence is obtained from these negative controls and the result is similar to that of the evidence sample, then the results will not be considered for comparison purposes. The mtDNA types should be determined for all laboratory personnel involved in the mtDNA analysis, and those sequences should be maintained on file. The positive control must yield the appropriate result.

All appropriate equipment should be calibrated periodically. Most manufacturers recommend appropriate schedules. Depending on the need or laboratory policy, the frequency of calibration may be increased. Instruments to calibrate include: thermal cycler, waterbath, heat block, CE, sequencer, refrigerators, and freezers.

AMPLIFICATION

The primers and primer sets used for mtDNA amplification are listed below. Mitochondrial DNA primers are named for the particular strand of mtDNA from which they are derived, in numerical form from the terminal 3′ base of the primer. The convention is L for light strand and H for heavy strand, followed by the number of that base as originally published (see Anderson et al., Reference 1). Primers should be diluted to 100 µM and aliquoted in small aliquots that should be stored at -20°C. A 10 µM working stock of small aliquots should be prepared for each primer and stored at -20°C.

For routine reference samples consisting of blood and/or saliva, and for unknown samples such as fresh bone, teeth, or forcibly removed hairs where a potentially abundant amount of mtDNA template is present in the sample, HV1 and HV2 can be amplified using the primers listed in Table 1.

Table 1. Primers
Hypervariable Region 1
A1: (L 15997) 5'-CAC CAT TAG CAC CCA AAG CT-3'
B2: (H 16236) 5'-CTT TGG AGT TGC AGT TGA TG-3'
A2: (L 16159) 5'-TAC TTG ACC ACC TGT AGT AC-3'
B1: (H 16391) 5'-GAG GAT GGT GGT CAA GGG AC-3'
Hypervariable Region 2
C1: (L 048) 5'-CTC ACG GGA GCT CTC CAT GC-3'
D2: (H 285) 5'-GGG GTT TGG TGG AAA TTT TTT TG-3'
C2: (L 172) 5'-ATT ATT TAT CGC ACC TAC GT-3'
D1: (H 408) 5'-CTG TTA AAA GTG CAT ACC GCC A-3'

For unknown samples, such as hairs without sheath tissue or degraded bone and teeth where little mtDNA template is expected, HV1A, HV1B, HV2A, and HV2B should be amplified using the primer sets listed in Table 2.

Table 2. Primer Set	
Region	**Primer Set**
HV1	A1, B1
HV2	C1, D1
HV1A	A1, B2
HV1B	A2, B1
HV2A	C1, D2
HV2B	C2, D1

Note: For mitochondrial DNA analysis, distilled deionized water is listed in the protocols. If contaminating DNA (above acceptable levels) is found due to the distilled, deionized water, then the use of high purity water (VWR Scientific Products, West Chester, PA, USA; Catalog No. 72060-0904) is recommended.

10.1 | AMPLIFICATION PROCEDURE

Materials and Reagents

- GeneAmp® PCR System 9600 Thermal Cycler (PE Biosystems, Foster City, CA, USA

- 10× PCR buffer
- Bovine serum albumin (BSA)
- dATP
- dTTP
- dCTP
- dGTP
- Primers (Table 1)
- *Taq* DNA polymerase
- Distilled, deionized water
- MicroAmp™ reaction tubes (PE Biosystems)
- Positive control DNA
- MicroAmp caps (PE Biosystems)
- MicroAmp retainer (PE Biosystems)
- Microcon™ 100 unit (Millipore, Bedford, MA, USA)

Procedure

1. Thermal cycling parameters for the GeneAmp PCR System 9600 thermal cycler are as follows:

   a. 95°C for 1 min

   b. Cycles:
      95°C for 10 s
      60°C for 30 s
      72°C for 30 s

   c. 15°C hold for 10 min

   In general, use 36 cycles for hair, bones, and teeth and 32 cycles for bodily fluids such as blood and saliva.

   If the slot blot hybridization assay (see Chapter 4, entitled Determination of the Quantity of DNA) shows the presence of at least 100 pg of genomic DNA, use 32 cycles and amplify HV1 and HV2 regardless of sample type.

2. If samples have been stored frozen, vortex-mix the sample tubes briefly, and pulse centrifuge in a microcentrifuge.

3. If the amount of extracted DNA is known through pre-amplification quantification analysis, amplification should be performed on 10 μL of an approximately 10 pg/μL solution of the extract.

4. Determine the number of samples to be amplified, including controls. Calculate the amount of reagents in each amplification master mixture using the following formulation:

Table 3. Amplification Master Mixture for 25 μL Reactions	
Distilled, deionized water	6.0 μL
10× PCR buffer	2.5 μL
Each dNTP (4)	0.5 μL
Each primer (30 μM) (2)	0.5 μL
BSA (1.6 μg/μL)	2.5 μL
Taq DNA polymerase (5 U/μL)	1.0 μL
DNA template (extract)	10.0 μL

5. Label MicroAmp reaction tubes with the appropriate sample designation. Label a tube for each positive and negative control. A positive and negative control should be prepared for each region amplified.

6. After the master mixture has been prepared, aliquot 15 μL into each MicroAmp reaction tube and place the reaction tubes in an amplification rack.

7. Add 10 μL of distilled deionized water to each negative control tube and cap each tube.

   Note: Pipet tips are changed between each sample.

8. Add 10 μL of the extract to the appropriate tube beginning with the reagent blanks and cap each tube.

9. Add 10 μL of positive control DNA (total 100 pg) to each positive control tube. Cap tubes.

10. Cap the MicroAmp reaction tubes with MicroAmp caps and carry the tubes to the thermal cycler.

11. A MicroAmp retainer should be maintained in the thermal cycler. Place the reaction tubes in the thermal cycler. Start the thermal cycler amplification program. Maintain a record of the heat block position of each tube.

12. The amplification program is completed in approximately 2 h. The samples should be stored at 4°C in a space dedicated to post-amplification processing.

Table 4. Volumes for Varying Numbers of Samples				
(All volumes are in µL)				
Component	1 Sample	12 Samples 14× Mixture	24 Samples 26× Mixture	36 Samples 38× Mixture
10× PCR buffer	2.5	35.0	65.0	95.0
BSA (1.6 µg/µL)	2.5	35.0	65.0	95.0
dATP	0.5	7.0	13.0	19.0
dCTP	0.5	7.0	13.0	19.0
dGTP	0.5	7.0	13.0	19.0
dTTP	0.5	7.0	13.0	19.0
primer (L) 30 µM	0.5	7.0	13.0	19.0
primer (H) 30 µM	0.5	7.0	13.0	19.0
Taq DNA polymerase 5 U/µL	1.0	14.0	26.0	38.0
Distilled deionized water	6.0	84.0	156.0	228.0
TOTALS	15.0	210.0	390.0	570.0

10.2 AMPLICON PURIFICATION

Procedure

1. Assemble and label a Microcon 100 unit for each sample. Prepare the Microcon 100 concentrators by adding 400 µL of distilled deionized water to the filter side of each concentrator.

2. Add the entire amplified product from each MicroAmp reaction tube to the appropriately labeled concentrator.

3. Centrifuge the Microcon 100 concentrators for 5 min at 3000× *g*.

   Note: additional centrifuge time may be required to filter the entire volume.

4. Discard the wash and return the filtrate cups to the concentrators.

5. Add 400 µL of filtered, distilled deionized water to the filter side of each Microcon 100 concentrator.

6. Centrifuge again at 3000 × *g* for 5 min and discard the filtrate cups.

7. Add 40 µL of filtered, distilled deionized water to the filter side of each Microcon 100 concentrator and place a retentate cup on the top of each concentrator.

8. Briefly vortex-mix the Microcon 100 concentrators with the retentate cups pointing upward.

9. Invert each concentrator with its retentate cup and centrifuge in a micro-centrifuge at 10,000× *g* for 3 min.

10. Discard the concentrators. Cap the retentate cups containing the purified amplicon for quantification and sequencing. Samples may be stored at 4°C.

10.3 POST-AMPLIFICATION QUANTIFICATION

Following post-amplification purification with Microcon 100 devices, the quantity of amplified mtDNA samples is determined by CE using the P/ACE 2050 or P/ACE 5000 instrument (Beckman Instruments, Fullerton, CA, USA) and a 50-μm High Resolution Gas Chromatography Capillary. The instrument is operated, and the data collected and analyzed by the System Gold™ Software (Version 8.0) (Beckman Instruments).

Materials and Reagents

- P/ACE 2050 or P/ACE 5000 instrument
- System Gold Software (Version 8.0)
- Soldering iron
- Cleaving stone

- Microcon 100 devices
- 50-μm High Resolution Gas Chromatography Capillary
- Sulfuric acid
- Methanol
- Distilled, deionized water
- Quantitation standard
- Sample vial
- Hydroxyethyl cellulose (HEC; Aldrich, Allentown, PA, USA)

Capillary Preparation

Procedure

1. Warm a soldering iron located in a fume hood.

2. Using a cleaving stone, cut the capillary to a length of 27 cm.

3. Measure approximately 6.5 cm from the right end of the capillary and mark it with a wide-tipped permanent marker. This segment should line up with the detector window of the exterior capillary cartridge. The etching or removal of the exterior coating of the capillary at this region will create a transparent window on the capillary. This region should be approximately 5 mm in length. Creating a larger region will weaken the capillary in this area, increasing the likelihood of breakage during assembly.

4. Working in a fume hood, place the marked region under the hot iron and carefully, using a plastic or Pasteur pipet, drop fuming sulfuric acid onto the tip of the iron. The fuming sulfuric acid will remove the capillary coating.

   CAUTION: Gloves, laboratory coat, and safety glasses should be worn when working with sulfuric acid.

5. Carefully assemble the capillary into the P/ACE System's 2050 (or 5000) capillary cartridge. Follow the instructions in the Capillary Replacement Procedure Manual (P/N 266910). Calibrate the laser-induced fluorescence (LIF) detector (Beckman Instruments) according to the LIF Detector Manual (Catalog No. 015-726017) using the Capillary Performance Test Kit (Catalog No. 3384070).

6. Load the cartridge into the P/ACE instrument following the instructions in the Beckman Users Manual.

7. The capillary should be rinsed with methanol before use.

For overnight storage, when not in use, the capillary ends should be immersed in distilled deionized water. Replacement of the capillary is not necessary unless breakage occurs or resolution decreases and rinsing does not improve resolution. It should be noted that the 1% HEC buffer solutions (Appendix 3) and water must be changed approximately every 40 samples. When not in use, buffer vials should be stored at room temperature in a manner that eliminates light exposure.

Programming System Gold

System Gold uses computer controlled programs referred to as methods. Enter or create a method for mtDNA quantification as described in the System Gold Users Manual. The method consists of the following steps.

Procedure

1. Fill the capillary with buffer for 2 min.

2. Rinse the inlet port in distilled deionized water for 6 s.

3. Inject the sample for 45 s with high pressure.

4. Separate for 4 min at constant voltage of 15 kV.

5. Rinse the capillary with methanol for 3 min.

   This method should be saved and used for all mtDNA quantification.

A second method should be created entitled Shutdown to rinse the capillary at the end of each sample set (see Users Manual). It should include a 5 to 10 min rinse of methanol and a 5 to 10 min rinse of distilled deionized water.

Table 5. The P/ACE HPCE [Base] Time Programming

Menu Is Set For:
1. Reinject: 0.
2. Channel: B.
3. EXT EVENTS: all 4 are OFF.
4. Wait for Temp: NO.
5. Prerinse No. 1: HIGH, FWD, 2.00 min, VIAL In 32, VIAL Out 10.
6. Prerinse No. 2: LOW, FWD, 0.01 min, VIAL In 33, VIAL Out 10.
7. Inject No. 2: 0.0 sec.
8. Inject No. 3: 0.0 sec. Table Line 1:
9. Time INIT: 0.0; Funct.: VOLT; Value: 15.0; Dur.: 0.17; In Vial: 34; Out Vial: 1. Table Line 2:
10. Time INIT: 4.0; Funct.: PRES; Value: HIGH; Dur.: 3.0; Dir.: FWD; In Vial 31; Out Vial: 10; Stop Data: YES Table Line 3:
11. Time INIT: 7.0; End: YES.

Table 6. The P/ACE Detector Time Program

Menu Is Set For:
1. Data Rate (Hz): 10.
2. Channel: A, LIF.
3. Rise Time: 1.0.
4. Chart Marks: OFF.
5. Negative Offset (%): 0.0.
6. Detector Polarity: NORMAL.
7. Range (AU): 0.0.
8. PMT Gain: 1.
9. Total Capillary Length (cm): 27.0.
10. Wavelengths Ex/Em (nm): 488 520.
11. Capillary to Detector (cm): 20.0. Table Line 1:
12. Time INIT: 0.5; Auto Zero: YES. Table Line 2:
13. Time INIT: 4.0; End: YES.

Sample Preparation

Procedure

1. Prior to sample preparation, turn on the laser and initiate a methanol/buffer rinse as described in the P/ACE System's 2050 (or 5000) Users Manual to rinse the column. This rinse should consist of 20 min with methanol followed by 20 min with buffer.

2. Soak and rinse all rubber caps in distilled deionized water and let air-dry.

3. Prepare a dilution of the quantification standard by mixing 2 μL of the quantification standard with 498 μL distilled deionized water. Lightly vortex-mix.

4. Pipet 1 μL of each purified mtDNA amplification product into the appropriate sample vial (P/ACE/eCAP vial P/N 727013).

5. Pipet 24 μL of the quantification standard dilution into each tube. Mix by pipetting up and down, making sure no bubbles are present. Alternatively, 2 to 5 μL of sample may be added, followed by the appropriate volume of the quantification standard up to a total of 25 μL.

6. Place the sample tubes into glass vials, each containing a spring insert. Place a rubber cap on each vial, making sure the incision in the cap remains closed.

7. Open the lid and place the glass vials into the appropriately numbered slots beginning with slot 11 in the inlet carousel of the CE instrument.

8. Place fresh 1% HEC buffer into the three buffer vials designated for rinsing and separating. Fill the vials up to the neck. Place the rubber caps on the vials and place in appropriate carousel positions.

9. Fill a vial with methanol and place a rubber cap on the vial. Place the methanol vial in the appropriate inlet carousel position.

10. Fill a vial with distilled deionized water and place a rubber cap on the vial. Place the vial in the slot designated for the water wait.

11. Place an empty vial with a rubber cap in the slot designated for waste in the outlet carousel. This vial is the common waste vial. Close the lid.

Sample Table Preparation

Procedure

1. Turn on the computer.
2. Pull down the Utilities menu and highlight Previous Screen.
3. Under the System 1 menu, highlight Edit Sample Table.
4. Under the column headed Method Name, type the name of the Method.
5. Under the column headed Sample Name, type the sample names (Q and

K designation). The names will be limited to 8 characters.

6. When all sample names have been entered, type Shutdown under the column headed by Method Name and Equ on the same line under the column headed by Type.

7. When the sample table is complete, pull down the Operations menu and highlight Save Table as. . .. The name should be the laboratory number. Press enter.

8. Click on OK. The computer will return to the sample table.

9. Click on the box to the right of Name. A window will appear with a list of sample table names. Click on the one just created. Click on OK.

10. Pull down Operations menu and highlight Print Table.

11. Pull down Operations menu and highlight Exit & Run Table. The method will begin operation.

12. All sample results should be stored on disk.

10.4 CALCULATION OF QUANTITY OF MITOCHONDRIAL DNA

The quantity of the amplified mtDNA is obtained by dividing the area of the amplified mtDNA peak by the area of the quantification standard peak. The quantification standard is added to every sample at a concentration of 384 pg/μL. The following equations are used:

Direct ratio of product peak to the quantification standard:

$$\frac{\text{Area mtDNA/Retention Time}}{\text{Area Quant. Stnd./Retention Time}} \times 9.6 \text{ ng/μL} = \text{ng mtDNA/μL}$$

Should the amount of amplified DNA as assessed by CE be less than 1 ng/μL, the quantity of amplified DNA should be determined again using more sample DNA. In such instances, the above quantitation procedure should be repeated, with either 2, 3, 4, or 5 μL of sample DNA, and 23, 22, 21, or 20 μL, respectively, of the quantification standard dilution. In such instances, the amount of amplified mtDNA/μL will be corrected to reflect the appropriate dilution. An example of such a calculation, using 3 μL of amplified sample, is shown below:

$$\frac{\text{Area mtDNA/Retention Time}}{\text{Area Quant. Stnd./Retention Time}} \times \frac{9.6 \text{ ng/μL } (22/24)}{3} = \text{ng mtDNA/μL}$$

10.5 SAMPLE PREPARATION FOR SEQUENCING

Materials and Reagents

- Model 377 automated DNA sequencer (PE Biosystems)
- GeneAmp 9600 thermal cycler

- Centri-Sep™ centrifuge columns (Princeton Separations, Adelphia, NJ, USA)
- ABI PRISM™ dRhodamine Terminator Cycle Sequencing Ready Reaction Kit (PE Biosystems)
- Single sequencing primer
- MicroAmp reaction tube
- pGEM
- Distilled, deionized water
- M13 primer
- MicroAmp caps

Centrifuge Column Preparation

Procedure

Prior to setting up the cycle sequencing reaction, Centri-Sep centrifuge columns are prepared to remove unincorporated fluorescent terminators from the completed sequencing reaction.

1. Gently tap the column to force the powder to the bottom of the column (the white filter is on the bottom).
2. Remove the top cap of the column and add 800 µL distilled deionized water.
3. Replace the cap, invert, vortex-mix at high speed, invert again, vortex-mix, and tap firmly until the gel solution is thoroughly mixed and any bubbles have risen to the top.
4. Allow the gel to hydrate a minimum of 30 min at room temperature. Columns can be stored at 4°C. Let columns equilibrate to room temperature before use.

Cycle Sequencing Reaction

Cycle sequencing is performed using the ABI PRISM dRhodamine Terminator Cycle Sequencing Ready Reaction Kit. The reaction is carried out by combining kit reagents, mtDNA template, and a single sequencing primer in a MicroAmp reaction tube. The reaction mixture is subjected to a series of controlled temperature changes in a GeneAmp PCR system 9600 thermal cycler (Figure 4). The optimum mtDNA template concentration for sequencing should be between 20 and 35 ng mtDNA/7 µL of added template. However, this range is not absolute, and lesser and greater amounts of template can be used if necessary. A minimum of 7 ng of amplified DNA template can be used for cycle sequencing.

The primers used for sequencing are identical to those used in amplification (see section entitled Amplification). The primers A4 and B4 are utilized on samples that contain a homopolymeric region (i.e., a C-stretch) from positions 16184 to 16194. They are designed to bind near the homopolymeric region and prime mtDNA synthesis on their respective strands. Primer A4 is

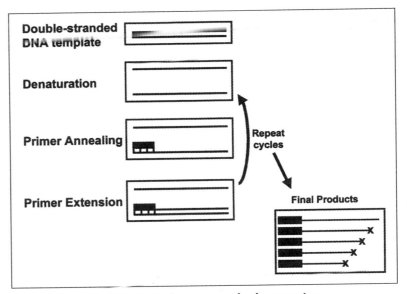

Figure 4. Schematic displaying the basic process of cycle sequencing.

used to sequence the HV1B region. Primer B4 is used to sequence the IIV1A region.

A4: (L 16209)   5′-CCC CAT GCT TAC AAG CAA GT-3′ (HVIB)
B4: (H 16164)   5′-TTT CAT GTC GAT TGG GTT T-3′ (HVIA)

Procedure

1. To prepare purified mtDNA template from the Microcon 100 retentate, dilute the retentate (based upon the quantity determined by CE quantification) to 20 to 35 ng mtDNA per 7 µL of added template. Label MicroAmp reaction tubes with the appropriate sample designation. Label a reaction tube for pGEM® (Promega, Madison, WI, USA).

2. Add 9.5 µL of sequencing ready reaction mixture to each tube.

3. Add 7 µL (approximately 20 to 35 ng) of the mtDNA template to the appropriate tubes.

4. Add 3.0 µL of pGEM control to the appropriately labeled tube.

5. Add 6.0 µL of distilled deionized water to the pGEM control.

6. Add 3.5 µL of the appropriate primer at a concentration of 1 µM to each sample tube.

7. Add 1.5 µL of the M13 primer to the pGEM control.

8. Place a strip of MicroAmp caps on each row of tubes. Seal the caps tightly.

9. Place the tubes into the GeneAmp 9600 thermal cycler, close and tighten the sample cover. Turn on the thermal cycler and program as follows:

a. 96°C for 1 min
b. 25 cycles:
    96°C for 15 s
    50°C for 1 s
    60°C for 1 min
c. 15°C hold for 10 min

Sample Preparation for Electrophoresis

Procedure

1. The excess distilled deionized water must be drained from the centrifuge columns. Remove the top cap, then the bottom cap, and let the column drain in a rack. If drainage does not begin immediately, apply air pressure to the top of the column with a pipet bulb.

2. Place the appropriate number of wash tubes in a centrifuge and place the drained centrifuge columns into the wash tubes.

3. Centrifuge at 1500× g for 2 min.

4. While the columns are centrifuging, label the collection tubes with the appropriate sample designation.

5. Remove the columns from the wash tubes and place them into the sample collection tubes. Discard the wash tubes.

6. When the cycle sequencing is complete, remove the entire reaction mixture from each reaction tube and carefully load it onto the top of the gel material. It is important to dispense the sample onto the center of the gel.

   Do not touch the sides of the column with the reaction mixture or pipet tip, and do not disturb the surface of the gel. This step is best accomplished by looking down into the tube rather than looking at it from the side.

7. Place the collection tubes with centrifuge columns in a centrifuge and centrifuge again at 1500× g for 2 min.

8. Discard the centrifuge columns.

9. Place the collection tubes containing purified sequencing fragments into a vacuum centrifuge and centrifuge to dryness. Do not overdry. This should take 15–20 min, depending on individual equipment.

10. Add water to wells in a heating block and set the temperature at 95°C.

10.6 SEQUENCING USING THE MODEL 377 DNA SEQUENCER

The procedure given here should be followed when using the Model 377 DNA Sequencer.

Materials and Reagents

- Gel casting cassette for 377
- Glass plates for 377 (36 cm long)
- Gel spacers
- Wonder Wedge (Amersham Pharmacia Biotech, Piscataway, NJ, USA)

- Long Ranger™ gel solution or Long Ranger ready-made Singel packs (FMC BioProducts, Rockland, ME, USA)
- 1× Tris-borate-EDTA (TBE) buffer
- 5% Alconox™
- 377 loading solution (50 µL deionized formamide plus 10 µL EDTA/blue dextran)

For acrylamide gels:
- 10% ammonium persulfate
- 40% acrylamide stock
- Urea
- Distilled, deionized water
- AG® 501-X8 (D) Resin (Bio-Rad, Hercules, CA, USA)
- 10× TBE buffer
- 1× TBE buffer
- 0.2 µm nylon Nalgene™ filter
- Tetramethylethylenediamine (TEMED)
- 377 loading solution (50 µL deionized formamide plus 10 µL EDTA/blue dextran)

10.7 SEQUENCING GEL PREPARATION

Procedure

1. Prior to each use, rinse both sides of the glass plates with hot water. The use of soap products should be kept to a minimum. However, if soap is used, rinse thoroughly with distilled water to remove any soap residue. Place the plates in an upright position and allow to air-dry (if methanol is used to decrease drying time, use only a high-purity HPLC/spectral grade methanol).

2. A 25-mL volume of gel solution is required for a 36-cm long gel. The Long Ranger ready-made packs contain sufficient gel solution (i.e., 50 mL) for two sequencing gels. Follow directions on outside of ready-made packs to prepare.

   If preparing a 5% Long Ranger gel from the 50% stock Gel Solution, combine the following:

   2.5 mL Long Ranger Gel Solution
   9.0 g urea
   2.5 mL 10× TBE buffer

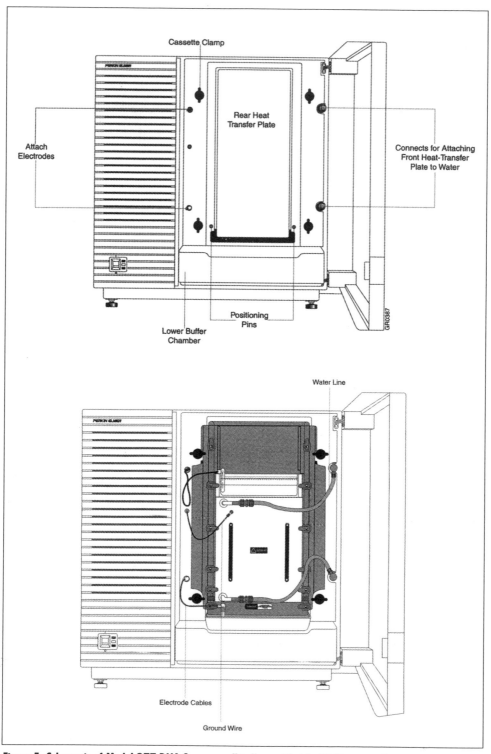

Figure 5. Schematic of Model 377 DNA Sequencer (kindly provided by PE Biosystems).

If preparing a 4.25% acrylamide gel, combine the following:

    2.65 mL 40% acrylamide stock solution

    9.0 g urea

    12.5 mL distilled deionized water

    0.25g AG 501-X8 (D) Resin

3. Stir the solution until the urea crystals have dissolved (approximately 5 min).

4. Prepare 10% ammonium persulfate (0.1 g ammonium persulfate plus 1 mL distilled deionized water).

5. While the solution stirs, prepare the cassette and plates.

6. Dry each plate gently with a lint-free tissue. The manufacturer's numbers etched on each plate should be on the outside. Determine which side of each plate is to be the inside (i.e., in contact with the gel).

7. Place the cassette on a bench top. Lift the beam stop at the bottom and set the plate clamps on each side to the open position.

8. Place the rear plate into the cassette. Make sure the inside of the plate is free of water droplets, dust, and lint.

9. Place the 0.2-mm gel spacers on each side with cut edges toward the top of the plate. Water droplets can be placed under the spacers to secure them to the plate.

10. Align the top plate (inside down) on top of the spacers. The bottom ends of the plates should be flush, and the notch of the front plate should be oriented toward the top of the cassette.

11. Push the plates toward the bottom of the cassette. Make certain the cassette stops are firmly against the upper edge of the rear plate notches. Push the plates from the top to ensure close contact at this surface.

12. Turn the plate clamps to lock the plates in position. Place the cassette on a level surface.

13. Finish preparing the gel solution.

For the Long Ranger gel mixture adjust the volume to 25 mL with distilled deionized water. Add 200 µL 10% ammonium persulfate and 20 µL TEMED. Swirl the solution.

For the acrylamide gel mixture add 2.5 mL 10× TBE buffer to the bottom portion of a 0.2 µm nylon Nalgene filter. Filter the gel solution, degas 2–5 min, and adjust the volume to 25 mL with distilled deionized water. Add 125 µL 10% ammonium persulfate and 17.5 µL TEMED to the gel solution. Swirl to mix and to initiate polymerization. Do not introduce bubbles.

14. Working quickly, draw the entire solution into a 25-mL pipet or syringe. While tapping on the top plate, dispense the solution against the notched edge. Move the pipet from one side to the other while steadily dispensing the solution. When the gel solution reaches the opposite side, position the pipet in the center and finish dispensing.

15. When the gel solution has completely filled all empty spaces and bub-

bles have been removed, insert the gel casting comb and place three binder clips on the top plate over the inner edge of the comb. Allow to polymerize for a minimum of 2 h at room temperature.

16. After 1 h, the Long Ranger gel should be covered with wet paper towels (soaked in 1× TBE buffer) to prevent drying. Wrap the Long Ranger gel in plastic wrap for overnight storage. Although polyacrylamide gels can be stored in a similar fashion, overnight storage is not recommended.

Note: Do not mark the glass plates with markers or pens; the ink may fluoresce and interfere with subsequent analysis.

10.8   CREATING THE SAMPLE SHEET AND RUN FILE

Procedure

1. While the gel is polymerizing, prepare the sample sheet and run file for data collection. Open the Model 377 Data Collection software.

2. To create a sample sheet, choose New from the File menu.

3. Click the applicable sample sheet icon (Sequence Sample) in the window that appears.

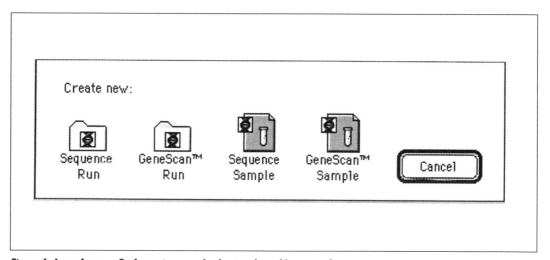

Figure 6. Icons for step 3 of creating sample sheet and run file protocol.

4. Enter appropriate information. Next, choose Save from the File menu. The dialog box that appears shows the default sample file name, but it can be changed. Click Save. Choose Close from the File menu or click the close box.

#	Sample Name	DyeSet/Primer	Matrix	Comments
1	PGEM	DT {dR Set Any-Prime▶	dRhod dyes f▶	Guam-2 9/29/99
2	FE-A	DT {dR Set Any-Prime▶	dRhod dyes f▶	See CE sheets for ng seq.
3	FE-B	DT {dR Set Any-Prime▶	dRhod dyes f▶	
4	FE-C	DT {dR Set Any-Prime▶	dRhod dyes f▶	
5	FE-D	DT {dR Set Any-Prime▶	dRhod dyes f▶	
6	NC-A	DT {dR Set Any-Prime▶	dRhod dyes f▶	
7	NC-B	DT {dR Set Any-Prime▶	dRhod dyes f▶	
8	NC-C	DT {dR Set Any-Prime▶	dRhod dyes f▶	
9	NC-D	DT {dR Set Any-Prime▶	dRhod dyes f▶	
10	217-A	DT {dR Set Any-Prime▶	dRhod dyes f▶	
11	217-B	DT {dR Set Any-Prime▶	dRhod dyes f▶	
12	217-C	DT {dR Set Any-Prime▶	dRhod dyes f▶	

Sample Sheet "SS GUAM-2 9/29/99 hsl"

Sequence Analysis Sample Sheet

Figure 7. Sample sheet for step 4 of creating sample sheet and run file protocol.

5. To create a run file, choose New from the File menu. Click the applicable run folder icon (Sequence Run) in the window that appears.

6. Choose the parameters for the run from the pop-up menus.

7. Choose modules to be used for the plate check (Plate Check A), the prerun module (PR2XA), and the run module (Run 2XA for Dye Terminator chemistry and 36E-1200 for dithodamine chemistry).

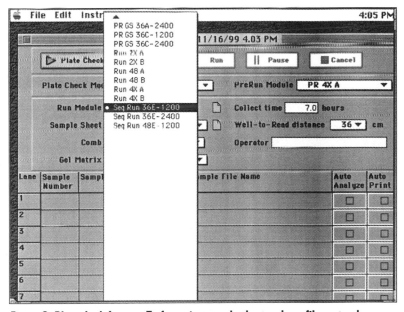

Figure 8. Plate check for step 7 of creating sample sheet and run file protocol.

8. Choose the sample sheet from the sample sheet pop-up menu.

Note: The sample sheet information cannot be edited in the Run window. Make any changes in the sample sheet and select the sample sheet again in the Run window to make the changes effective.

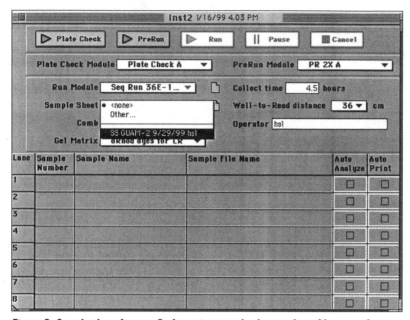

Figure 9. Sample sheet for step 8 of creating sample sheet and run file protocol.

9. Choose the type of comb and the well-to-read distance (e.g., 36 cm).

10. Choose the appropriate matrix file based on the chemistry used (e.g., 96051543 matrix for the Dye Terminator kit and dRhod matrix for dRhodamine Terminator chemistry).

11. Enter the length of the run (a minimum of 4 h is needed, although the run times can be extended) in the Collect Time entry field by clicking open, entering value, and saving.

12. Enter user's initials in the Operator entry field. Click Save from the File menu.

13. When finished completing the information in the run file, do not close the window. It can be used to perform the plate check.

10.9   INSTRUMENT SETUP

Procedure

1. Prepare 2 L of 1× TBE buffer to be used as the gel running buffer (200 mL 10× TBE buffer plus 1800 mL distilled deionized water).

2. Turn on the sequencer and allow to warm up.

3. After 2 h have elapsed, remove the binder clips from the plate.

4. Remove the plates from the cassette and rinse with warm distilled deionized water. When the area is wet, remove the comb. Remove any pieces of acrylamide from the well region.

5. Dry the plates with lint-free tissues. Be careful not to smear any residual acrylamide across the plates. Reload the plates into the cassette.

6. Open the door of the sequencer. To allow proper positioning of the cassette, make certain the rear heat transfer plate and two small positioning pins at the bottom of the chamber are clean.

7. Place the lower buffer chamber in the bottom shelf of the electrophoresis chamber.

8. To mount the cassette, place the cassette against the rear transfer plate. The side spacers on the gel assembly that are located in the region of the rear plate notches should press against the two small positioning pins at the bottom of the chamber. The bottom of the plates should extend into the lower buffer chamber. Close the clamps to secure the cassette.

9. Next, scan the plates before adding buffer and loading samples to ensure that no peaks are produced by fluorescent particles on the glass plates or in the gel.

10. Click Plate Check in the run window. Observe the scan window that appears.

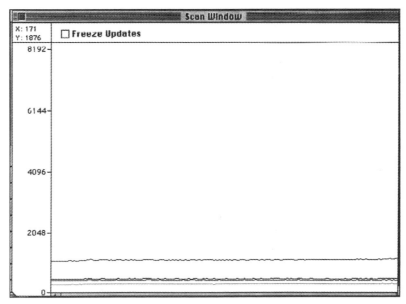

Figure 10. Scan Window for step 10 of instrument setup protocol.

11. The scan window should show a relatively flat line across the screen in each of the four colors. If there are no peaks on the scan go to step 15. If there are any peaks on the scan (possibly due to dust on the glass), open the door of the sequencer to pause the plate check.

12. Remove the cassette from the sequencer. Reclean both sides of the glass in the laser read region with distilled deionized water and a lint-free tissue. Replace the cassette in the electrophoresis chamber. Close the door and the scan will resume.

13. If peaks are still observed after cleaning the plates, the gel mixture may contain particles. The scan window indicates which channels contain contaminating particles. (Use Table 3-1 on pg. 3-13 in the Model 377 DNA Sequencing User's Manual to determine which lanes should not be used.)

14. If fluorescent signal of any color is above 2000 relative fluorescent units (RFU), open the Gel Image window (under the Window menu) and observe the general appearance of the gel background. Various substances that fluoresce brightly may not burn off during the Prerun. A high fluorescent background may obscure signal from a weak sample.

15. If the plates appear clean, cancel the plate check and open the door of the sequencer. Place the upper buffer chamber against the top of the glass plates. Make sure the overhang lip of the buffer chamber rests on the top of the glass plates. Close the top plate clamps to secure it into position.

16. Fill the upper buffer chamber with 1× TBE buffer to the leveling plastic piece on the left hand side of the buffer chamber. This should take approximately 580 to 600 mL of buffer. Check for leaks. Insert the lid to prevent evaporation.

17. If a leak is observed, siphon off the buffer and clean the gasket area. Clean the area on the plate that contacts the gasket and reassemble.

18. Fill the lower buffer chamber with 1× TBE buffer to the top edge of the dam that separates the chamber from the overflow reservoir in front of it.

19. Attach the front heat transfer plate. Open the plate clamps between the upper buffer chamber and the beam stop. Push the front heat transfer plate against the front of the cassette so that it rests on the platform that is part of the beam stop bar.

20. Secure it with the plate clamps. Attach the quick-connect water lines and the ground cable. Plug in the electrode cables extending from the buffer chambers.

21. Prerun the gel to ensure the system is working correctly. Click the Prerun button in the run window. The prerun is set for one hour. Open the Status option under the Window menu. Allow gel temperature to warm to 48°C before inserting the sharks-tooth comb and loading the gel.

22. Prepare a small amount of 377 loading solution (50 μL deionized formamide plus 10 μL EDTA/blue dextran). Once a gel temperature of 48°C is obtained, rinse the well. Pipet the loading solution between the

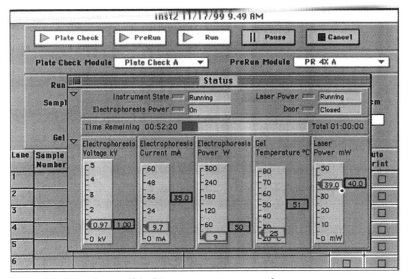

Figure 11. Status for step 21 of instrument setup protocol.

plates while slowly dragging the pipet from one end of the well to the other. Prerun for 3 min. This will allow the loading buffer to cover the surface of the gel, thus allowing for better visualization of the gel surface when inserting the comb.

23. Open the door to pause the Prerun. Using the loading solution as a guide for the surface of the gel, insert the sharks-tooth comb. Carefully align the center registration line on the comb with the registration mark in the notch of the glass plate. Slide the comb between the plates. The teeth should slightly indent the surface. Do not remove the comb once the teeth have penetrated the gel surface. This will cause sample to leak into adjacent wells. Rinse wells, close the door, and allow the Prerun to continue.

10.10   SAMPLE PREPARATION AND GEL LOADING

Procedure

1. Prepare the DNA samples. Resuspend the DNA with 5 μL loading solution (5:1 deionized formamide:25 mm EDTA with 50 mg/mL blue dextran). Vortex-mix, then briefly centrifuge down.

2. To ensure that the sequencer accurately defines the lanes, alternate samples are loaded and subjected to electrophoresis for 2 min before the remaining samples are loaded. The odd lanes are loaded first, followed by the even lanes. Negative control or reagent blank samples should not be placed in the first lane of the gel since this lane is important for lane tracking and should contain a sample.

3. Place the samples into the wells of the heating block (95°C) for 2 min and then place on ice.

4. Open the door of the sequencer to pause the Prerun. Flush out the wells with a transfer pipet or syringe. Pipet 1.5 µL of each odd-numbered sample into the appropriate odd-numbered wells. Close the door to allow the prerun to resume. Subject to electrophoresis for 2 min to allow the samples to enter the gel.

5. After 2 min, flush all the wells to remove any residual formamide from previously loaded wells. Pipet 1.5 µL of each even-numbered sample into the appropriate even-numbered wells.

6. Click Cancel in the run window to cancel the Prerun.

7. Click Run to start the run.

8. A dialog box appears, allowing you to change the name and location of the gel file created by the run. When satisfied with the name and location, click OK. The run will now begin.

10.11   CLEANING THE PLATES, COMB, AND SPACERS

Procedure

1. When the run is completed, the sequencer can be shut off.

2. Siphon the buffer out of the upper and lower buffer chambers. Remove the heat plate. Take the cassette out of the sequencer.

3. Remove the plates from the cassette. Gently insert a Wonder Wedge into the edge of the bottom of the plates and pry the plates apart. Do not exert too much force or the plates may chip. Remove the well comb and the spacers.

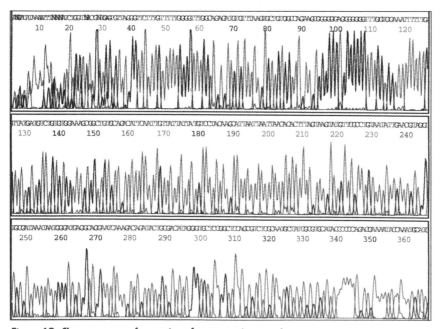

Figure 12. Chromatogram of a portion of a sequencing reaction.

4. Lay a large paper towel on the gel. Remove the gel from the plate by lifting the paper towel off the plate. Discard in an appropriate waste container.

5. Rinse the plates, comb, and spacers with 5% Alconox solution. Use a gloved hand to dislodge any gel bits. Allow plates to air dry.

6. Rinse the buffer chambers with distilled deionized water and allow to air dry.

10.12 POST-ELECTROPHORESIS ANALYSIS

Procedure

1. Click open Run Folder (this should launch the Sequencing Analysis software).

2. Click open the gel file.

3. Review gel file.

10.13 PROTOCOL FOR SEQUENCE ANALYSIS: SEQUENCE NAVIGATOR SOFTWARE

Procedure

1. Open Sequence Navigator Software (PE Biosystems).

2. Open New Layout (under File).

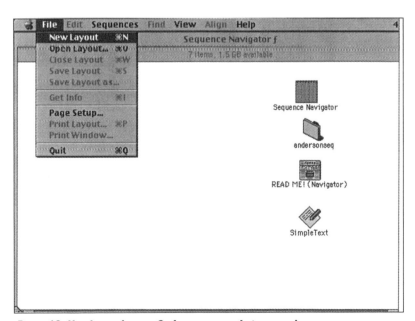

Figure 13. New layout for step 2 of sequence analysis protocol.

3. Open Import Sequences (under Sequences).

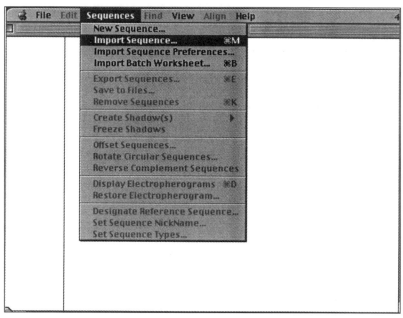

Figure 14. Import Sequences for step 3 of sequence analysis protocol.

4. Open desired file and select and import desired file.

5. In a similar fashion import the matching reverse sequence and corresponding reference sequence (also known as Anderson sequence).

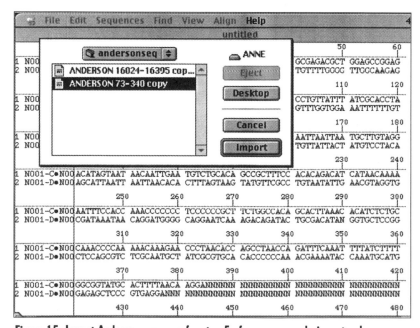

Figure 15. Import Anderson sequence for step 5 of sequence analysis protocol.

6. Select the name of the sequence beginning with the reverse primer and open Reverse Complement Sequence (under Sequences).

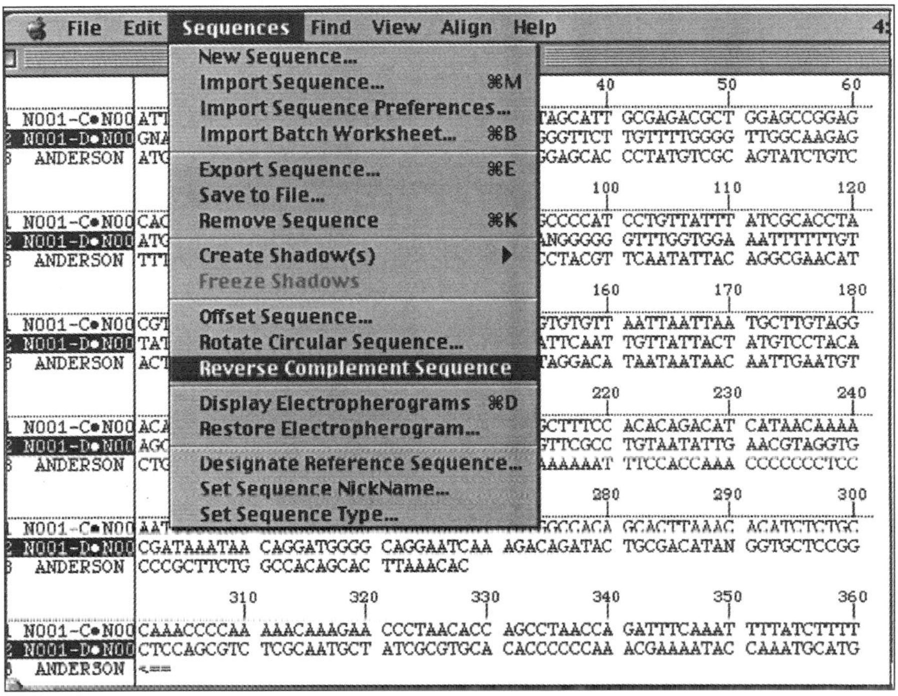

Figure 16. Reverse complement sequence for step 6 of sequence analysis protocol.

7. Locate the desired starting point of each sequence and delete the preceding bases:

    HVI begins with TTCTTT for primers A1, B1, and B2.
    HVII begins with GTGCAC for primers C1, D1, and D2.

    Then, delete bases designated N at the end of the sequences.

8. Select the two sequences to compare and open Create Shadow(s)—Compare Two Sequences (under Sequences).

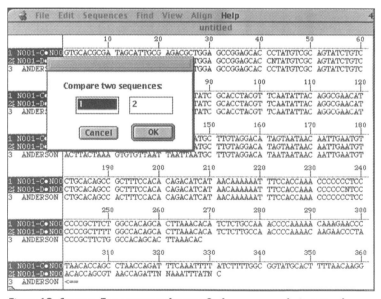

Figure 17. Compare Two Sequences for step 8 of sequence analysis protocol.

9. Confirm the sequences selected and then click OK in the dialog box.

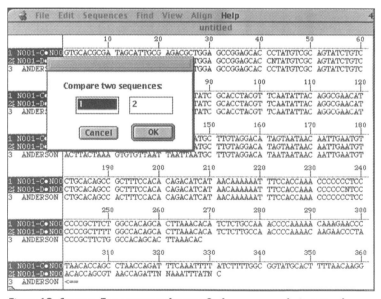

Figure 18. Compare Two sequences for step 9 of sequence analysis protocol.

10. Highlight the region of the sequence of interest and open Display Electropherogram (under Sequences); edit the sequence region and then close; repeat this analysis for all sequences that need editing.

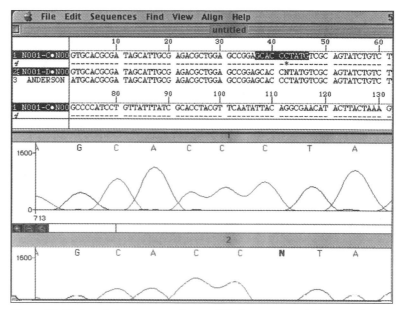

Figure 19. Display electropherogram for step 10 of sequence analysis protocol.

11. After editing, remove shadow sequence and realign all sequences according to Anderson reference by opening Offset Sequences (under Sequences).

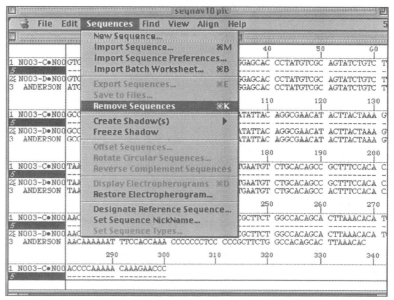

Figure 20. Remove sequences for step 11 of sequence analysis protocol.

12. In the dialog box, click on Offset Sequence(s) to the right and enter desired offset number:

For HVI offset by 16023,
For HVII offset by 72.

Then click OK.

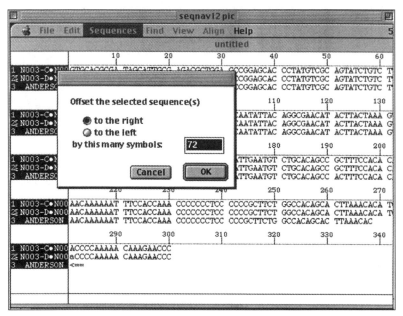

Figure 21. Offset the selected sequences for step 12 of sequence analysis protocol.

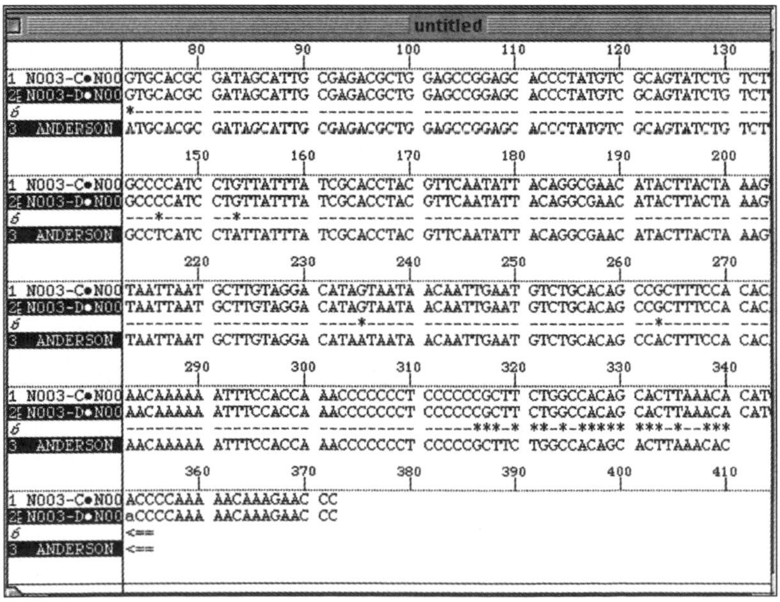

Figure 22. Displayed sequence compared with Anderson sequence.

13. To return the sequences to the beginning of the layout format, click on Home (under View).

14. To compare the sequence(s) to Anderson reference sequence, open Create Shadow(s)—Compare Two sequences (save this shadow sequence in the final layout file).

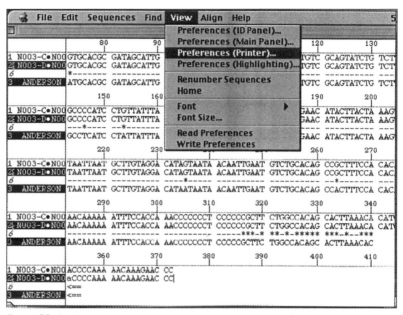

Figure 23. Preference for step 16 of sequence analysis protocol.

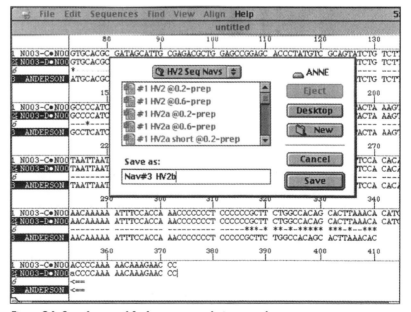

Figure 24. Save for step 16 of sequence analysis protocol.

15. Check each identified polymorphic site (i.e., a difference in sequence from the Anderson reference) by reviewing the electropherogram at each region.

16. Name the layout using Preferences (Printer) (under View) and save under File using Save Layout as....

SUGGESTED READING

1. Anderson, S., A.T. Bankier, B.G. Barrell, M.H.L. de Bruijn, A.R. Coulson, I.C. Drouin, I.C. Eperon, D.P. Nierlick, et al. 1981. Sequence and organization of the mitochondrial genome. Nature *290*:457-465.

2. Aquadro, C.F. and B.D. Greenberg. 1983. Human mitochondrial DNA variation and evolution: analysis of nucleotide sequences from seven individuals. Genetics *103*:287-312.

3. Bodenhagen, D. and D.A. Clayton. 1974. The number of mitochondrial deoxyribonucleic acid genomes in mouse L and human HeLa cells. J. Biol. Chem. *249*:7791.

4. Brown, W.M., M. George, and A.C. Wilson. 1979. Rapid evolution of mitochondrial DNA. Proc. Natl. Acad. Sci. USA *76*:1967-1971.

5. Budowle, B., D.E. Adams, C.T. Comey, and C.R. Merril. 1990. Mitochondrial DNA: a possible genetic material suitable for forensic analysis, p. 76-97. *In* H.C. Lee and R.E. Gaensslen (Eds.). Advances in Forensic Sciences. Year Book Medical Publishers, Chicago.

6. Budowle, B., M.R. Wilson, J.A. DiZinno, C. Stauffer, M.A. Fasano, M.M. Holland, and K.L. Monson. 1999. Mitochondrial DNA regions HVI and HVII population data. Forensic Sci. Int. *103*:23-35.

7. Butler, J.M., R.O. Allen, B.R. McCord, J.M. Jung, M.R. Wilson, and B. Budowle. 1994. Quantitation of PCR products by capillary electrophoresis using laser fluorescence. J. Chromatogr. *658*:271-280.

8. Case, J.T. and D.C. Wallace. 1981. Maternal inheritance of mitochondrial DNA polymorphisms in cultured human fibroblasts. Somat. Cell Genet. 7:103-108.

9. Clayton, D.A. 1992. Structure and function of the mitochondrial genome. J. Inher. Metab. Dis. *15*:439-447.

10. DeGiorgi, C. and C. Saccone. 1989. Mitochondrial genome in animal cells: structure, organization, and evolution. Cell Biophys. *14*:67-77.

11. Giles, R.E., H. Blanc, H.M. Cann, and D.C. Wallace. 1980. Maternal inheritance of human mitochondrial DNA. Proc. Natl. Acad. Sci. USA *77*:6715-6719.

12. Ginther, C., L. Issel-Tarver, and M.C. King. 1992. Identifying individuals by sequencing mitochondrial DNA from teeth. Nat. Genet. *2*:135-138.

13. Goodall, D.M., S.J. Williams, and D.K. Lloyd. 1991. Quantitative aspects of capillary electrophoresis. Trends Anal. Chem. *10*:272-279.

14. Gray, M.W. 1989. Origin and evolution of mitochondrial DNA. Annu. Rev. Cell Biol. *5*:25-50.

15. Greenberg, B.D., J.E. Newbold, and A. Sugino. 1983. Intraspecific nucleotide sequence variability surrounding the origin of replication in human mitochondrial DNA. Gene *21*:33-49.

16. Hauswirth, W.W., C.D. Dickel, D.J. Rowold, and M.A. Hauswirth. 1994. Inter- and intrapopulation studies of ancient remains. Experientia *50*:585-591.

17. Hayashi, J.I., M. Takemitsu, Y. Goto, and I. Nonaka. 1994. Human mitochondria and mitochondrial genome function as a single dynamic cellular unit. J. Cell Biol. *125*:43-50.

18. Higuchi, R.G., C.H. von Beroldigen, G.F. Sensabaugh, and H.A. Ehrlich. 1988. DNA typing from single hairs. Nature *332*:543-546.

19. Holland, M.M., D.L. Fisher, L.G. Mitchell, W.C. Rodriguez, J.J. Canik, C.R. Merril, and V.W. Weedn. 1993. Mitochondrial DNA sequence analysis of human skeletal remains: identification of remains from the Vietnam War. J. Forensic Sci. *38*:542-553.

20. Hopgood, R., K.M. Sullivan, and P. Gill. 1992. Strategies for automated sequencing of human mitochondrial DNA directly from PCR products. BioTechniques *13*:82-92.

21. Horai, S., K. Hayasaka, R. Kondo, K. Tsugane, and N. Tatakahata. 1995. Recent African origin of modern humans revealed by complete sequences of hominoid mitochondrial DNAs. Proc. Natl. Acad. Sci. USA *92*:532-536.

22. Hutchinson, C.A., J.E. Newbold, S.S. Potter, and M.H. Edgell. 1974. Maternal inheritance of mammalian mitochondrial DNA. Nature *251*:536-538.

23. Lestienne, P. and N. Bataille. 1994. Mitochondrial DNA alterations and genetic diseases: a review. Biomed. Pharmacother. *48*:199-214.

24. McCord, B.R., D.M. McClure, and J.M. Jung. 1993. Capillary electrophoresis of PCR-amplified DNA using fluorescence detection with an intercalating dye. J. Chromatogr. *652*:75-82.

25. McCord, B.R., J.M. Jung, and E.A. Holleran. 1993. High resolution capillary electrophoresis of

forensic DNA using a non-gel sieving buffer. J. Chromatogr, *16*;1963-1981

26. Merriwether, D.A., F. Rothhammer, and R.E. Ferrell. 1994. Genetic variation in the New World: ancient teeth, bone, and tissue as sources of DNA. Experientia *50*:592-601.

27. Monnat, R.J. and L.A. Loeb. 1985. Nucleotide sequence preservation of human mitochondrial DNA. Proc. Natl. Acad. Sci. USA *82*:2895-2899.

28. Monnat, R.J., C.L. Maxwell, and L.A. Loeb. 1985. Nucleotide sequence preservation of human leukemic mitochondrial DNA. Cancer Res. *45*:1809-1814.

29. Monnat, R.J. and D.T. Reay. 1986. Nucleotide sequence identity of mitochondrial DNA from different human tissues. Gene *43*:205-211.

30. Orrego, C. and M.C. King. 1990. Determination of familial relationships, p. 416-426. *In* M.A. Innis, D.H. Gelfand, J.J. Sninsky, and T.J. White (Eds.), PCR Protocols: A Guide to Methods and Applications. Academic Press, San Diego.

31. Paabo, S., J.A. Gifford, and A.C. Wilson. 1988. Mitochondrial DNA sequences from a 7000-year old brain. Nucleic Acids Res. *16*:9775-9778.

32. Piercy, R., K.M. Sullivan, N. Benson, and P. Gill. 1993. The application of mitochondrial DNA typing to the study of white Caucasian genetic identification. Int. J. Legal Med. *106*:85-90.

33. Prober, J.M., G.L. Trainor, R.J. Dam, F.W. Hobbs, C.E. Robertson, R.J. Zagursky, A.J. Cocuzza, M.A. Jensen, and K. Baumeister. 1987. A system for rapid DNA sequencing with fluorescent chain-terminating dideoxynucleotides. Science *238*:336-341.

34. Saccone, C. 1994. The evolution of mitochondrial DNA. Curr. Opin. Genet. Develop. *4*:875-881.

35. Sanger, F., S. Nicklen, and A.R. Coulson. 1977. DNA sequencing with chain-terminating inhibitors. Proc. Natl. Acad. Sci. USA *74*:5463-5468.

36. Smith, L.M., J.Z. Sanders, R.J. Kaiser, P. Hughes, C. Dodd, C.R. Connell, C. Heiner, S.B.H. Kent, and L.E. Hood. 1986. Fluorescence detection in automated DNA sequence analysis. Nature *321*:674-679.

37. Stoneking, M. 1994. Mitochondrial DNA and human evolution. J. Bioenerg. Biomembr. *26*:251-259.

38. Stoneking, M., D. Hedgecock, R.G. Higuchi, L. Vigilant, and H.A. Ehrlich. 1991. Population variation of human mtDNA control region sequences detected by enzymatic amplification and sequence specific oligonucleotide probes. Am. J. Hum. Genet. *48*:370-382.

39. Sullivan, K.M., R. Hopgood, and P. Gill. 1992. Identification of human remains by amplification and automated sequencing of mitochondrial DNA. Int. J. Legal Med. *105*:83-86.

40. Sullivan, K.M., R. Hopgood, B. Lang, and P. Gill. 1991. Automated amplification and sequencing of human mitochondrial DNA. Electrophoresis *12*:17-21.

41. Torroni, A., M.T. Lott, M.F. Cabell, Y.S. Chen, L. Lavergne, and D. Wallace. 1994. MtDNA and the origin of Caucasians: identification of ancient Caucasian specific haplogroups, one of which is prone to a recurrent somatic duplication in the D-loop region. Am. J. Hum. Genet. *55*:760-776.

42. Tritschler, H.J. and R. Medori. 1993. Mitochondrial DNA alterations as a source of human disorders. Neurology *43*:280-288.

43. Vigilant, L., R. Pennington, H. Harpending, T.D. Kocher, and A.C. Wilson. 1989. Mitochondrial DNA sequences in single hairs from a southern African population. Proc. Natl. Acad. Sci. USA *86*:9350-9354.

44. Vigilant, L., M. Stoneking, H. Harpending, K. Hawkes, and A.C. Wilson. 1991. African populations and the evolution of human mitochondrial DNA. Science *253*:1503-1507.

45. Wallace, D.C. 1994. Mitochondrial DNA sequence variation in human evolution and disease. Proc. Natl. Acad. Sci. USA *91*:8739-8746.

46. Ward, R.H., B.L. Frazier, K. Dew-Jager, and S. Paabo. 1991. Extensive mitochondrial diversity within a single Amerindian tribe. Proc. Natl. Acad. Sci. USA *88*:8720-8724.

47. Wilson, A.C. and R.L. Cann. 1992. The recent African genesis of humans. Sci. Am. *4*:68-73.

48. Wilson, M.R., J.A. DiZinno, D. Polanksey, J. Replogle, and B. Budowle. 1995. Validation of mitochondrial DNA sequencing for forensic casework analysis. Int. J. Legal Med.*108*:68-74.

49. Wilson, M.R., D. Polanskey, J. Butler, J.A. DiZinno, J. Replogle, and B. Budowle. 1995. Extraction, PCR amplification, and sequencing of mitochondrial DNA from human hair shafts. BioTechniques *18*:662-669.

50. Wilson, M.R., D. Polanskey, J. Replogle, J.A. DiZinno, and B. Budowle. 1997. A family exhibiting heteroplasmy in the human mitochondrial DNA control region reveals both somatic mosaicism and pronounced segregation of mitotypes. Hum. Genet. *100*:167-171.

51. Wilson, M.R., M. Stoneking, M.M. Holland, J.A. DiZinno, and B. Budowle. 1993. Guidelines for the use of mitochondrial DNA sequencing in forensic science. Crime Lab. Digest *20*:68-77.

52. Wrischnik, L.A., R. Higuchi, M. Stoneking, and H.A. Erlich. 1987. Length mutations in human mitochondrial DNA: direct sequencing of enzymatically amplified DNA. Nucleic Acids Res. *15*:529-542.

Appendix I: Quality Assurance Standards

Laboratory quality assurance (QA) is the documented verification that proper procedures have been carried out by skilled and highly trained personnel in such a way that valid and reliable results can be obtained. The validity of DNA typing results centers on correctly identifying true non-matches and matches from DNA profile comparisons. When the same methods are used, test reliability implies reproducibility under defined conditions of use and should transcend different laboratories and practitioners. Quality assurance must encompass all significant aspects of the DNA typing process including: personnel education and training, documentation of records, data analysis, quality control of reagents and equipment, technical controls, proficiency testing, reporting of results, and auditing of the laboratory procedures. Quality assurance and appropriate standards for DNA typing evolved initially from the experience of clinical laboratories but more recently have been carefully defined through consensus from forensic laboratories represented in the North American non-regulatory initiative, the Technical Working Group on DNA Analysis Methods (TWGDAM). It is important to note that quality assurance guidelines in forensic science must retain enough flexibility to accommodate the nature of forensic samples as well as future advancements in recombinant DNA technology and molecular biology. Good quality management is paramount for obtaining quality results.

The DNA Identification Act of 1994 contained within the Omnibus Crime Control Act (The Violent Crime Control and Law Enforcement Act of 1994, Public Law 103-322, 108 stat. 1796) expressly authorizes the FBI Director to establish QA standards for laboratories performing forensic DNA testing. Through the deliberations and recommendations of a DNA Advisory Board, the Director of the FBI has issued the following QA standards, which are required for performing DNA typing in a forensic laboratory.

DNA Typing Protocols: Molecular Biology and Forensic Analysis
By B. Budowle, J. Smith, T. Moretti, J. DiZinno
©2000 Eaton Publishing, Natick, MA

Quality Assurance Standards
for Forensic DNA Testing Laboratories

Preface

Throughout its deliberation concerning these quality standards, the DNA Advisory Board recognized the need for a mechanism to ensure compliance with the standards. An underlying premise for these discussions was that accreditation would be required to demonstrate compliance with the standards and therefore assure quality control and a quality program. Accordingly, the Board recommends that forensic laboratories performing DNA analysis seek such accreditation with all deliberate speed. Additionally, the Board strongly encourages the accrediting bodies to begin positioning themselves to accommodate the increasing demand for accreditation.

Proposed Mechanism to Recommend Changes to Standards

Once the Director of the FBI has issued standards for quality assurance for forensic DNA testing, the DNA Advisory Board may recommend revisions to such standards to the FBI Director, as necessary. In the event that the duration of the DNA Advisory Board is extended beyond March 10, 2000 by the FBI Director, the Board may continue to recommend revisions to such standards to the FBI Director. In the event that the DNA Advisory Board is not extended by the FBI Director after March 10, 2000, the Technical Working Group on DNA Analysis Methods [TWGDAM] may recommend revisions to such standards to the FBI Director, as necessary.

Effective Date

These standards shall take effect October 1, 1998.

INTRODUCTION

This document consists of definitions and standards. The standards are quality assurance measures that place specific requirements on the laboratory. Equivalent measures not outlined in this document may also meet the standard if determined sufficient through an accreditation process.

REFERENCES:

American Society of Crime Laboratory Directors-Laboratory Accreditation Board (ASCLD- LAB), *ASCLD-LAB Accreditation Manual*, January 1994, and January, 1997.

International Standards Organization (ISO)/International Electrotechnical Commission (IEC), *ISO/IEC Guide* 25-1990, (1990) American National Standards Institute, New York, NY.

Technical Working Group on DNA Analysis Methods, "Guidelines for a Quality Assurance Program for DNA Analysis," *Crime Laboratory Digest*, April 1995, Volume 22, Number 2, pp. 21-43.

42 Code of Federal Regulations, Chapter IV (10-1-95 Edition), Health Care Financing Administration, Health and Human Services.

1. SCOPE

The standards describe the quality assurance requirements that a laboratory, which is defined as a facility in which forensic DNA testing is performed, should follow to ensure the quality and integrity of the data and competency of the laboratory. These standards do not preclude the participation of a laboratory, by itself or in collaboration with others, in research and development, on procedures that have not yet been validated.

2. DEFINITIONS

As used in these standards, the following terms shall have the meanings specified:

(a) Administrative review is an evaluation of the report and supporting documentation for consistency with laboratory policies and for editorial correctness.

(b) Amplification blank control consists of only amplification reagents without the addition of sample DNA. This control is used to detect DNA contamination of the amplification reagents.

(c) Analytical procedure is an orderly step by step procedure designed to ensure operational uniformity and to minimize analytical drift.

(d) Audit is an inspection used to evaluate, confirm, or verify activity related to quality.

(e) Calibration is the set of operations which establish, under specified conditions, the relationship between values indicated by a measuring instrument or measuring system, or values represented by a material, and the corresponding known values of a measurement.

(f) Critical reagents are determined by empirical studies or routine practice to require testing on established samples before use on evidentiary samples in order to prevent unnecessary loss of sample.

(g) Commercial test kit is a pre-assembled kit that allows the user to conduct a specific forensic DNA test.

(h) Examiner/analyst is an individual who conducts and/or directs the analysis of forensic casework samples, interprets data and reaches conclusions.

(i)   Forensic DNA testing is the identification and evaluation of biological evidence in criminal matters using DNA technologies.

(j)   Known samples are biological material whose identity or type is established.

(k)   Laboratory is a facility in which forensic DNA testing is performed.

(l)   Laboratory support personnel are individual(s) who perform laboratory duties and do not analyze evidence samples.

(m)   NIST is the National Institute of Standards and Technology.

(n)   Polymerase Chain Reaction (PCR) is an enzymatic process by which a specific region of DNA is replicated during repetitive cycles which consist of (1) denaturation of the template; (2) annealing of primers to complementary sequences at an empirically determined temperature; and (3) extension of the bound primers by a DNA polymerase.

(o)   Proficiency test sample is biological material whose DNA type has been previously characterized and which is used to monitor the quality performance of a laboratory or an individual.

(p)   Proficiency testing is a quality assurance measure used to monitor performance and identify areas in which improvement may be needed. Proficiency tests may be classified as:

   (1)   Internal proficiency test is one prepared and administered by the laboratory.

   (2)   External proficiency test, which may be open or blind, is one which is obtained from a second agency.

(q)   Qualifying test measures proficiency in both technical skills and knowledge.

(r)   Quality assurance includes the systematic actions necessary to demonstrate that a product or service meets specified requirements for quality.

(s)   Quality manual is a document stating the quality policy, quality system and quality practices of an organization.

(t)   Quality system is the organizational structure, responsibilities, procedures, processes and resources for implementing quality management.

(u)   Reagent blank control consists of all reagents used in the test process without any sample. This is to be used to detect DNA contamination of the analytical reagents.

(v)   Reference material (certified or standard) is a material for which values are certified by a technically valid procedure and accompanied by or traceable to a certificate or other documentation which is issued by a certifying body.

(w)   Restriction Fragment Length Polymorphism (RFLP) is generated by cleavage by a specific restriction enzyme and the variation is due to restriction site polymorphism and/or the number of different repeats contained within the fragments.

(x)   Review is an evaluation of documentation to check for consistency, accuracy, and completeness.

(y)   Second agency is an entity or organization external to and independent of the laboratory and which performs forensic DNA analysis.

(z)   Secure area is a locked space (for example, cabinet, vault or room) with access restricted to authorized personnel.

(aa) Subcontractor is an individual or entity having a transactional relationship with a laboratory.

(bb) Technical manager or leader (or equivalent position or title as designated by the laboratory system) is the individual who is accountable for the technical operations of the laboratory.

(cc) Technical review is an evaluation of reports, notes, data, and other documents to ensure an appropriate and sufficient basis for the scientific conclusions. This review is conducted by a second qualified individual.

(dd) Technician is an individual who performs analytical techniques on evidence samples under the supervision of a qualified examiner/analyst and/or performs DNA analysis on samples for inclusion in a database. Technicians do not evaluate or reach conclusions on typing results or prepare final reports.

(ee) Traceability is the property of a result of a measurement whereby it can be related to appropriate standards, generally international or national standards, through an unbroken chain of comparisons.

(ff) Validation is a process by which a procedure is evaluated to determine its efficacy and reliability for forensic casework analysis and includes:

   (1)   Developmental validation is the acquisition of test data and determination of conditions and limitations of a new or novel DNA methodology for use on forensic samples.

   (2)   Internal validation is an accumulation of test data within the laboratory to demonstrate that established methods and procedures perform as expected in the laboratory.

3. QUALITY ASSURANCE PROGRAM

STANDARD   3.1   **The laboratory shall establish and maintain a documented quality system that is appropriate to the testing activities.**

3.1.1 The quality manual shall address at a minimum:
(a)   Goals and objectives
(b)   Organization and management
(c)   Personnel Qualifications and Training
(d)   Facilities

(e)   Evidence control
(f)   Validation
(g)   Analytical procedures
(h)   Calibration and maintenance
(i)   Proficiency testing
(j)   Corrective action
(k)   Reports
(l)   Review
(m)   Safety
(n)   Audits

4. ORGANIZATION AND MANAGEMENT

STANDARDS   4.1   The laboratory shall:

(a)   have a managerial staff with the authority and resources needed to discharge their duties and meet the requirements of the standards in this document.

(b)   have a technical manager or leader who is accountable for the technical operations.

(c)   specify and document the responsibility, authority, and interrelation of all personnel who manage, perform or verify work affecting the validity of the DNA analysis.

5. PERSONNEL

STANDARD   5.1   Laboratory personnel shall have the education, training and experience commensurate with the examination and testimony provided. The laboratory shall:

5.1.1 have a written job description for personnel to include responsibilities, duties and skills.

5.1.2 have a documented training program for qualifying all technical laboratory personnel.

5.1.3 have a documented program to ensure technical qualifications are maintained through continuing education.

5.1.3.1 Continuing education - the technical manager or leader and examiner/analyst(s) must stay abreast of developments within the field of DNA typing by reading current scientific literature and by attending seminars, courses, professional meetings or documented training sessions/classes in relevant subject areas at least once a year.

5.1.4 maintain records on the relevant qualifications, training, skills and experience of the technical personnel.

5.2 The technical manager or leader shall have the following:

5.2.1 <u>Degree requirements</u>: The technical manager or leader of a laboratory shall have at a minimum a Master's degree in biology-, chemistry- or forensic science-related area and successfully completed a minimum of 12 semester or equivalent credit hours of a combination of undergraduate and graduate course work covering the subject areas of biochemistry, genetics and molecular biology (molecular genetics, recombinant DNA technology), or other subjects which provide a basic understanding of the foundation of forensic DNA analysis as well as statistics and/or population genetics as it applies to forensic DNA analysis.

5.2.1.1 The degree requirements of section 5.2.1 may be waived by the American Society of Crime Laboratory Directors (ASCLD) or other organization designated by the Director of the FBI in accordance with criteria approved by the Director of the FBI. This waiver shall be available for a period of two years from the effective date of these standards. The waiver shall be permanent and portable.

5.2.2 <u>Experience requirements</u>: A technical manager or leader of a laboratory must have a minimum of three years of forensic DNA laboratory experience.

5.2.3 <u>Duty requirements</u>:

5.2.3.1 <u>General</u>: manages the technical operations of the laboratory.

5.2.3.2 <u>Specific duties</u>

(a) Is responsible for evaluating all methods used by the laboratory and for proposing new or modified analytical procedures to be used by examiners.

(b) Is responsible for technical problem

solving of analytical methods and for the oversight of training, quality assurance, safety and proficiency testing in the laboratory.

5.2.3.3 The technical manager or leader shall be accessible to the laboratory to provide onsite, telephone or electronic consultation as needed.

5.3   Examiner/analyst shall have:

5.3.1 at a minimum a BA/BS degree or its equivalent degree in biology-, chemistry- or forensic science- related area and must have successfully completed college course work (graduate or undergraduate level) covering the subject areas of biochemistry, genetics and molecular biology (molecular genetics, recombinant DNA technology) or other subjects which provide a basic understanding of the foundation of forensic DNA analysis, as well as course work and/or training in statistics and population genetics as it applies to forensic DNA analysis.

5.3.2 a minimum of six (6) months of forensic DNA laboratory experience, including the successful analysis of a range of samples typically encountered in forensic case work prior to independent case work analysis using DNA technology.

5.3.3 successfully completed a qualifying test before beginning independent casework responsibilities.

5.4   Technician shall have:

5.4.1 On the job training specific to their job function(s).

5.4.2 successfully completed a qualifying test before participating in forensic DNA typing responsibilities.

5.5   Laboratory support personnel shall have:

5.5.1 training, education and experience commensurate with their responsibilities as outlined in their job description.

6. FACILITIES

STANDARD 6.1 The laboratory shall have a facility that is designed to provide adequate security and minimize contamination. The laboratory shall ensure that:

6.1.1 Access to the laboratory is controlled and limited.

6.1.2 Prior to PCR amplification, evidence examinations, DNA extractions, and PCR setup are conducted at separate times or in separate spaces.

6.1.3 Amplified DNA product is generated, processed and maintained in a room(s) separate from the evidence examination, DNA extractions and PCR setup areas.

6.1.4 The laboratory follows written procedures for monitoring, cleaning and decontaminating facilities and equipment.

7. EVIDENCE CONTROL

STANDARD 7.1 The laboratory shall have and follow a documented evidence control system to ensure the integrity of physical evidence. This system shall ensure that:

7.1.1 Evidence is marked for identification.

7.1.2 Chain of custody for all evidence is maintained.

7.1.3 The laboratory follows documented procedures that minimize loss, contamination, and/or deleterious change of evidence.

7.1.4 The laboratory has secure areas for evidence storage.

STANDARD 7.2 Where possible, the laboratory shall retain or return a portion of the evidence sample or extract.

7.2.1 The laboratory shall have a procedure requiring that evidence sample/extract(s) are stored in a manner that minimizes degradation.

8. VALIDATION

STANDARD 8.1 The laboratory shall use validated methods and procedures for forensic casework analyses.

8.1.1 Developmental validation that is conducted shall be appropriately documented.

8.1.2 Novel forensic DNA methodologies shall undergo developmental validation to ensure the accuracy, precision and reproducibility of the procedure. The developmental validation shall include the following:

8.1.2.1 Documentation exists and is available which defines and characterizes the locus.

8.1.2.2 Species specificity, sensitivity, stability and mixture studies are conducted.

8.1.2.3 Population distribution data are documented and available.

8.1.2.3.1 The population distribution data would include the allele and genotype distributions for the locus or loci obtained from relevant populations. Where appropriate, databases should be tested for independence expectations.

8.1.3 Internal validation shall be performed and documented by the laboratory.

8.1.3.1 The procedure shall be tested using known and non- probative evidence samples. The laboratory shall monitor and document the reproducibility and precision of the procedure using human DNA control(s).

8.1.3.2 The laboratory shall establish and document match criteria based on empirical data.

8.1.3.3 Before the introduction of a procedure into forensic casework, the analyst or examination team shall successfully complete a qualifying test.

8.1.3.4 Material modifications made to analytical procedures shall be documented and subject to validation testing.

8.1.4 Where methods are not specified, the laboratory shall, wherever possible, select methods that have been published by reputable technical organizations or in relevant scientific texts or journals, or have been appropriately evaluated for a specific or unique application.

9. ANALYTICAL PROCEDURES

STANDARD 9.1 **The laboratory shall have and follow written analytical procedures approved by the laboratory management/technical manager.**

9.1.1 The laboratory shall have a standard operating protocol for each analytical technique used.

9.1.2 The procedures shall include reagents, sample preparation, extraction, equipment, and controls which are standard for DNA analysis and data interpretation.

9.1.3 The laboratory shall have a procedure for differential extraction of stains that potentially contain semen.

STANDARD 9.2 **The laboratory shall use reagents that are suitable for the methods employed.**

9.2.1 The laboratory shall have written procedures for documenting commercial supplies and for the formulation of reagents.

9.2.2 Reagents shall be labeled with the identity of the reagent, the date of preparation or expiration, and the identity of the individual preparing the reagent.

9.2.3 The laboratory shall identify critical reagents and evaluate them prior to use in casework. These critical reagents include but are not limited to:

(a) Restriction enzyme

(b) Commercial kits for performing genetic typing

(c) Agarose for analytical RFLP gels

(d) Membranes for Southern blotting

(e) K562 DNA or other human DNA controls

(f) Molecular weight markers used as RFLP sizing standards

(g) Primer sets

(h) Thermostable DNA polymerase

STANDARD 9.3 **The laboratory shall have and follow a procedure for evaluating the quantity of the human DNA in the sample where possible.**

9.3.1 For casework RFLP samples, the presence of high molecular weight DNA should be determined.

STANDARD 9.4 **The laboratory shall monitor the analytical procedures using appropriate controls and standards.**

9.4.1 The following controls shall be used in RFLP casework analysis:

9.4.1.1 Quantitation standards for estimating the amount of DNA recovered by extraction.

9.4.1.2 K562 as a human DNA control. (In monitoring sizing data, a statistical quality control method for K562 cell line shall be maintained.)

9.4.1.3 Molecular weight size markers to bracket known and evidence samples.

9.4.1.4 Procedure to monitor the completeness of restriction enzyme digestion.

9.4.2 The following controls shall be used for PCR casework analysis:

9.4.2.1 Quantitation standards which estimate the amount of human nuclear DNA recovered by extraction.

9.4.2.2 Positive and negative amplification controls.

9.4.2.3 Reagent blanks.

9.4.2.4 Allelic ladders and/or internal size makers for variable number tandem repeat sequence PCR based systems.

STANDARD 9.5 **The laboratory shall check its DNA procedures annually or whenever substantial changes are made to the protocol(s) against an appropriate and available NIST standard reference material or standard traceable to a NIST standard.**

STANDARD 9.6 **The laboratory shall have and follow written general guidelines for the interpretation of data.**

9.6.1 The laboratory shall verify that all control results are within established tolerance limits.

9.6.2 Where appropriate, visual matches shall be

supported by a numerical match criterion.

9.6.3 For a given population(s) and/or hypothesis of relatedness, the statistical interpretation shall be made following the recommendations 4.1, 4.2 or 4.3 as deemed applicable of the National Research Council report entitled "The Evaluation of Forensic DNA Evidence" (1996) and/or court directed method. These calculations shall be derived from a documented population database appropriate for the calculation.

10. EQUIPMENT CALIBRATION AND MAINTENANCE

STANDARD   10.1 The laboratory shall use equipment suitable for the methods employed.

STANDARD   10.2 The laboratory shall have a documented program for calibration of instruments and equipment.

10.2.1 Where available and appropriate, standards traceable to national or international standards shall be used for the calibration.

10.2.1.1 Where traceability to national standards of measurement is not applicable, the laboratory shall provide satisfactory evidence of correlation of results.

10.2.2 The frequency of the calibration shall be documented for each instrument requiring calibration. Such documentation shall be retained in accordance with applicable Federal or state law.

STANDARD   10.3 The laboratory shall have and follow a documented program to ensure that instruments and equipment are properly maintained.

10.3.1 New instruments and equipment, or instruments and equipment that have undergone repair or maintenance, shall be calibrated before being used in casework analysis.

10.3.2 Written records or logs shall be maintained for maintenance service performed on instruments and equipment. Such documentation shall be retained in accordance with applicable Federal or state law.

11. REPORTS

STANDARD   11.1 The laboratory shall have and follow written procedures for taking and maintaining case notes to support the conclusions drawn in laboratory reports.

    11.1.1   The laboratory shall maintain, in a case record, all documentation generated by examiners related to case analyses.

    11.1.2   Reports according to written guidelines shall include:

    (a) Case identifier
    (b) Description of evidence examined
    (c) A description of the methodology
    (d) Locus
    (e) Results and/or conclusions
    (f) An interpretative statement (either quantitative or qualitative)
    (g) Date issued
    (h) Disposition of evidence
    (i) A signature and title, or equivalent identification, of the person(s) accepting responsibility for the content of the report.

    11.1.3   The laboratory shall have written procedures for the release of case report information.

12. REVIEW

STANDARD   12.1 The laboratory shall conduct administrative and technical reviews of all case files and reports to ensure conclusions and supporting data are reasonable and within the constraints of scientific knowledge.

    12.1.1   The laboratory shall have a mechanism in place to address unresolved discrepant conclusions between analysts and reviewer(s).

STANDARD   12.2 The laboratory shall have and follow a program that documents the annual monitoring of the testimony of each examiner.

13. PROFICIENCY TESTING

STANDARD 13.1 Examiners and other personnel designated by the technical manager or leader who are actively engaged in DNA analysis shall undergo, at regular intervals of not to exceed 180 days, external proficiency testing in accordance with these standards. Such external proficiency testing shall be an open proficiency testing program.

13.1.1 The laboratory shall maintain the following records for proficiency tests:

(a) The test set identifier.
(b) Identity of the examiner.
(c) Date of analysis and completion.
(d) Copies of all data and notes supporting the conclusions.
(e) The proficiency test results.
(f) Any discrepancies noted.
(g) Corrective actions taken.

Such documentation shall be retained in accordance with applicable Federal or state law.

13.1.2 The laboratory shall establish at a minimum the following criteria for evaluation of proficiency tests:

(a) All reported inclusions are correct or incorrect.
(b) All reported exclusions are correct or incorrect.
(c) All reported genotypes and/or phenotypes are correct or incorrect according to consensus genotypes/phenotypes or within established empirically determined ranges.
(d) All results reported as inconclusive or uninterpretable are consistent with written laboratory guidelines. The basis for inconclusive interpretations in proficiency tests must be documented.
(e) All discrepancies/errors and subsequent corrective actions must be documented.
(f) All final reports are graded as satisfactory or unsatisfactory. A satisfactory grade is attained when

there are no analytical errors for the DNA profile typing data. Administrative errors shall be documented and corrective actions taken to minimize the error in the future.

(g) All proficiency test participants shall be informed of the final test results.

14. CORRECTIVE ACTION

STANDARD   **14.1 The laboratory shall establish and follow procedures for corrective action whenever proficiency testing discrepancies and/or casework errors are detected.**

14.1.1   The laboratory shall maintain documentation for the corrective action. Such documentation shall be retained in accordance with applicable Federal or state law.

15. AUDITS

STANDARD   **15.1 The laboratory shall conduct audits annually in accordance with the standards outlined herein.**

15.1.1   Audit procedures shall address at a minimum:

(a)   Quality assurance program
(b)   Organization and management
(c)   Personnel
(d)   Facilities
(e)   Evidence control
(f)   Validation
(g)   Analytical procedures
(h)   Calibration and maintenance
(i)   Proficiency testing
(j)   Corrective action
(k)   Reports
(l)   Review
(m)   Safety
(n)   Previous audits

15.1.2   The laboratory shall retain all documentation pertaining to audits in accordance with relevant legal and agency requirements.

STANDARD   15.2   Once every two years, a second agency shall participate in the annual audit.

16. SAFETY

STANDARD   16.1   The laboratory shall have and follow a documented environmental health and safety program.

17. SUBCONTRACTOR OF ANALYTICAL TESTING FOR WHICH VALIDATED PROCEDURES EXIST

STANDARD   17.1   A laboratory operating under the scope of these standards will require certification of compliance with these standards when a subcontractor performs forensic DNA analyses for the laboratory.

         17.1.1   The laboratory will establish and use appropriate review procedures to verify the integrity of the data received from the subcontractor.

Appendix II: Allele Frequency Data

GENERAL INFORMATION ON FREQUENCY CALCULATIONS

Allele frequency data are used to estimate the rarity of a single or multiple locus DNA profile. The frequencies are derived by the gene counting method from typing results derived by analyzing a sample population. The attached tables list allele frequencies for a few representative population groups.

For the RFLP/VNTR loci, alleles cannot be resolved into discrete entities as is done with PCR-based systems. Thus, a fixed bin method was devised to categorize alleles for practical application. All fragment lengths in each population sample are sorted into 31 fixed bin categories based on the size of fragments in the molecular weight marker. The number of DNA fragments that fell into each bin is divided by the total number of alleles (i.e., twice the number of individuals) in the sample population to determine the frequency of each bin. Then, the frequencies are rebinned whereby bins with fewer than five counts were merged with contiguous bins. After the bin tables are thus established, the frequency of an observed allele is estimated by determining in which bin(s) the fragment could reside, using a ±2.5% (or ±5.0%) measurement error window. If the measurement error window spans a bin boundary, the frequency of the higher frequency bin is assigned to the allele. A single-locus frequency of a two-band pattern is calculated using $2\ p_i p_j$, where p_i and p_j are the estimated binned allele frequencies for each VNTR band, while the frequency of occurrence of a single-band pattern is estimated using $2p$. The frequency of occurrence of a profile composed of multiple single-locus profiles is calculated as the product of the single-locus frequencies.

For PCR-based loci, allele frequencies are displayed without the convention of binning. Although of little impact on DNA profile frequency estimates, the notable differences in practice for calculating profile frequencies

DNA Typing Protocols: Molecular Biology and Forensic Analysis
By B. Budowle, J. Smith, T. Moretti, J. DiZinno
©2000 Eaton Publishing, Natick, MA

is the application of θ (an inbreeding coefficient). Thus, generally $p^2 + p(1\text{-}p)\,\theta$ is used to estimate the frequency of a homozygote profile instead of p^2 (or instead of 2p for VNTR loci). The $2p_i2p_j$ formula is still used to estimate heterozygous frequencies. A conservative value of 0.01 has been assigned as θ for the US population, and for some small, isolated populations, a value of 0.03 may be more appropriate.

Highly polymorphic loci will contain a large number of alleles, and some of the alleles could be rare or undetected in a sampling. To allay concerns that estimates presented might underestimate the frequency of occurrence of DNA profiles, some procedure should be used to compensate for sparse sampling of infrequent alleles in population databases. For the VNTR loci detected by RFLP typing, the fixed bin method is used to classify quasi-continuous distributions of alleles. To allow for use of small-size databases and to provide a bound on rare allele frequencies, a minimum allele frequency of rare alleles is derived by the rebinning process whereby bins with fewer than five counts are merged with contiguous bins. For PCR-based loci more discrete allelic data can be obtained than is possible with VNTRs typed by RFLP analysis. A minimum allele frequency for discretized allelic data can be estimated by using the formula 5/2N, where N is the sample size of the database. More sophisticated approaches exist, but this formula is simple and easy to calculate.

For most situations, allele frequency data from general population groups (i.e., African American, Caucasian, Hispanic, etc.) are sufficient. The data presented are examples of general population group distributions for each of the loci described in this book (excluding mtDNA sequence profiles, where the reference sequence for regions HVI and HVII is presented). Other population groups can be found in the literature.

POPULATION DATA - EXAMPLES

D1S7 Locus

Bin	Range (bp)	African American (n=718)	Caucasian (n=1190)	SW Hispanic (n=432)	SE Hispanic (n=610)	Chinese (n=250)
1	0-639	.000	.000	.000	.000	0.000
2	640-772	.000	.000	.000	.002	0.000
3	773-871	.000	.001	.000	.002	0.000
4	872-963	.000	.000	.002	.003	0.000
5	964-1077	.000	.002	.000	.002	0.000
6	1078-1196	.007	.002	.007	.003	0.004
7	1197-1352	.010	.006	.005	.013	0.004
8	1353-1507	.011	.009	.009	.013	0.004
9	1508-1637	.007	.012	.009	.018	0.012
10	1638-1788	.017	.011	.012	.015	0.012
11	1789-1924	.021	.013	.014	.020	0.016
12	1925-2088	.011	.010	.019	.016	0.032
13	2089-2351	.035	.029	.044	.031	0.040
14	2352-2522	.022	.020	.016	.010	0.036
15	2523-2692	.029	.013	.028	.041	0.032
16	2693-2862	.029	.029	.019	.016	0.032
17	2863-3033	.025	.031	.042	.021	0.032
18	3034-3329	.045	.046	.079	.028	0.028
19	3330-3674	.063	.068	.060	.056	0.052
20	3675-3979	.065	.055	.051	.056	0.060
21	3980-4323	.074	.062	.083	.090	0.076
22	4324-4821	.050	.067	.081	.044	0.080
23	4822-5219	.050	.055	.063	.044	0.052
24	5220-5685	.049	.060	.072	.056	0.064
25	5686-6368	.065	.063	.051	.064	0.056
26	6369-7241	.064	.079	.081	.075	0.080
27	7242-8452	.060	.077	.058	.084	0.064
28	8453-10093	.074	.076	.044	.075	0.048
29	10094-11368	.017	.033	.005	.025	0.048
30	11369-12829	.028	.019	.014	.026	0.004
31	12830-	.072	.051	.035	.051	0.052

n = number of alleles

D2S44 Locus

Bin	Range (bp)	African American (n=950)	Caucasian (n=1584)	SW Hispanic (n=430)	SE Hispanic (n=600)	Chinese (n=250)
1	0-639	.002	.000	.000	.000	.000
2	640-772	.055	.003	.002	.000	.000
3	773-871	.017	.003	.012	.000	.016
4	872-963	.013	.003	.042	.008	.028
5	964-1077	.021	.015	.028	.013	.028
6	1078-1196	.076	.024	.014	.018	.020
7	1197-1352	.092	.046	.070	.057	.032
8	1353-1507	.069	.035	.119	.068	.140
9	1508-1637	.086	.124	.112	.105	.120
10	1638-1788	.093	.107	.095	.105	.204
11	1789-1924	.084	.083	.100	.075	.080
12	1925-2088	.047	.050	.037	.048	.100
13	2089-2351	.076	.083	.077	.067	.072
14	2352-2522	.035	.038	.040	.018	.044
15	2523-2692	.028	.041	.028	.040	.056
16	2693-2862	.008	.040	.035	.045	.028
17	2863-3033	.032	.086	.040	.098	.004
18	3034-3329	.031	.089	.070	.098	.008
19	3330-3674	.039	.075	.051	.062	.004
20	3675-3979	.034	.023	.009	.035	.008
21	3980-4323	.019	.017	.012	.018	.004
22	4324-4821	.012	.006	.005	.007	.004
23	4822-5219	.004	.001	.000	.003	.000
24	5220-5685	.011	.001	.000	.003	.000
25	5686-6368	.013	.006	.002	.005	.000
26	6369-7241	.005	.001	.002	.002	.000
27	7242-8452	.000	.000	.000	.000	.000
28	8453-10093	.000	.000	.000	.000	.000
29	10094-11368	.000	.001	.000	.000	.000
30	11369-12829	.000	.000	.000	.000	.000
31	12830-	.000	.000	.000	.000	.000

n = number of alleles

D4S139 Locus

Bin	Range (bp)	African American (n=896)	Caucasian (n=1188)	SW Hispanic (n=422)	SE Hispanic (n=622)	Chinese (n=250)
1	0-639	.000	.000	.000	.000	.000
2	640-772	.000	.000	.000	.000	.000
3	773-871	.000	.000	.000	.000	.000
4	872-963	.000	.000	.000	.000	.000
5	964-1077	.000	.000	.000	.000	.000
6	1078-1196	.000	.000	.000	.000	.000
7	1197-1352	.000	.000	.000	.000	.000
8	1353-1507	.000	.000	.002	.000	.000
9	1508-1637	.000	.000	.002	.000	.000
10	1638-1788	.000	.000	.000	.000	.000
11	1789-1924	.000	.000	.000	.000	.000
12	1925-2088	.001	.000	.000	.000	.000
13	2089-2351	.068	.004	.007	.005	.000
14	2352-2522	.004	.000	.000	.002	.000
15	2523-2692	.011	.010	.005	.005	.000
16	2693-2862	.016	.003	.000	.011	.004
17	2863-3033	.012	.003	.002	.006	.016
18	3034-3329	.021	.014	.005	.013	.020
19	3330-3674	.052	.031	.014	.024	.037
20	3675-3979	.063	.023	.043	.023	.037
21	3980-4323	.077	.040	.028	.045	.069
22	4324-4821	.064	.047	.028	.061	.094
23	4822-5219	.066	.054	.059	.063	.098
24	5220-5685	.081	.072	.104	.095	.134
25	5686-6368	.084	.108	.137	.106	.122
26	6369-7241	.103	.191	.152	.167	.171
27	7242-8452	.109	.131	.175	.130	.098
28	8453-10093	.077	.095	.130	.101	.061
29	10094-11368	.029	.036	.021	.040	.016
30	11369-12829	.023	.035	.024	.032	.016
31	12830-	.038	.102	.059	.071	.008

n = number of alleles

D10S28 Locus

Bin	Range (bp)	African American (n=576)	Caucasian (n=858)	SW Hispanic (n=420)	SE Hispanic (n=460)	Chinese (n=424)
1	0-639	.000	.000	.000	.000	0.000
2	640-772	.002	.000	.002	.002	0.009
3	773-871	.000	.001	.002	.004	0.009
4	872-963	.005	.014	.014	.015	0.057
5	964-1077	.026	.051	.040	.041	0.090
6	1078-1196	.043	.044	.069	.041	0.054
7	1197-1352	.028	.017	.048	.035	0.057
8	1353-1507	.057	.040	.098	.072	0.085
9	1508-1637	.043	.078	.083	.093	0.050
10	1638-1788	.043	.087	.105	.096	0.045
11	1789-1924	.082	.083	.071	.091	0.026
12	1925-2088	.056	.047	.026	.046	0.054
13	2089-2351	.076	.059	.093	.078	0.057
14	2352-2522	.049	.019	.045	.035	0.038
15	2523-2692	.035	.016	.014	.024	0.064
16	2693-2862	.040	.042	.021	.020	0.087
17	2863-3033	.049	.049	.026	.037	0.057
18	3034-3329	.049	.048	.045	.039	0.038
19	3330-3674	.056	.065	.038	.059	0.035
20	3675-3979	.036	.045	.031	.039	0.031
21	3980-4323	.052	.072	.083	.046	0.038
22	4324-4821	.056	.068	.019	.054	0.014
23	4822-5219	.026	.014	.007	.011	0.005
24	5220-5685	.014	.007	.005	.004	0.002
25	5686-6368	.017	.027	.012	.009	0.000
26	6369-7241	.007	.006	.000	.002	0.000
27	7242-8452	.035	.000	.000	.007	0.000
28	8453-10093	.009	.000	.000	.000	0.000
29	10094-11368	.003	.000	.000	.000	0.000
30	11369-12829	.007	.000	.000	.000	0.000
31	12830-	.000	.000	.000	.000	0.000

n = number of alleles

D17S79 Locus

Bin	Range (bp)	African American (n=1100)	Caucasian (n=1552)	SW Hispanic (n=414)	SE Hispanic (n=628)	Chinese (n=252)
1	0-639	.001	.010	.005	.010	.000
2	640-772	.001	.003	.005	.000	.000
3	773-871	.005	.007	.010	.005	.000
4	872-963	.002	.001	.002	.003	.000
5	964-1077	.030	.003	.007	.003	.000
6	1078-1196	.034	.015	.007	.008	.016
7	1197-1352	.256	.224	.268	.232	.270
8	1353-1507	.195	.198	.135	.166	.222
9	1508-1637	.112	.263	.203	.255	.202
10	1638-1788	.107	.199	.109	.183	.063
11	1789-1924	.069	.028	.135	.083	.060
12	1925-2088	.076	.032	.087	.027	.127
13	2089-2351	.061	.010	.022	.013	.032
14	2352-2522	.012	.003	.002	.006	.008
15	2523-2692	.006	.002	.000	.000	.000
16	2693-2862	.012	.001	.002	.000	.000
17	2863-3033	.003	.000	.000	.003	.000
18	3034-3329	.009	.000	.000	.002	.000
19	3330-3674	.010	.001	.000	.002	.000
20	3675-3979	.000	.000	.000	.000	.000
21	3980-4323	.000	.000	.000	.000	.000
22	4324-4821	.000	.000	.000	.000	.000
23	4822-5219	.000	.000	.000	.000	.000
24	5220-5685	.000	.000	.000	.000	.000
25	5686-6368	.000	.000	.000	.000	.000
26	6369-7241	.000	.000	.000	.000	.000
27	7242-8452	.000	.000	.000	.000	.000
28	8453-10093	.000	.000	.000	.000	.000
29	10094-11368	.000	.000	.000	.000	.000
30	11369-12829	.000	.000	.000	.000	.000
31	12830-	.000	.000	.000	.000	.000

n = number of alleles

HLA-DQA1 ALLELE FREQUENCIES

ALLELE	AFRICAN AMERICAN (N=338)	CAUCASIAN (N=298)	SOUTHEASTERN HISPANIC (N=265)	SOUTHWESTERN HISPANIC (N=164)
1.1	0.129	0.122	0.149	0.134
1.2	0.281	0.178	0.175	0.137
1.3	0.049	0.064	0.083	0.043
2	0.108	0.133	0.157	0.091
3	0.092	0.183	0.181	0.213
4	0.342	0.320	0.255	0.381

N = number of individuals typed

PM ALLELE FREQUENCIES

PM loci observed allele frequencies in 145 unrelated African Americans					
Allele	LDLR	GYPA	HBGG	D7S8	Gc
A	0.224	0.479	0.507	0.614	0.103
B	0.776	0.521	0.197	0.386	0.707
C	NA	NA	0.297	NA	0.190

PM loci observed allele frequencies in 148 unrelated United States Caucasians					
Allele	LDLR	GYPA	HBGG	D7S8	Gc
A	0.453	0.584	0.470	0.615	0.257
B	0.547	0.416	0.524	0.385	0.172
C	NA	NA	0.007	NA	0.571

PM loci observed allele frequencies in 94 unrelated Southeastern Hispanics					
Allele	LDLR	GYPA	HBGG	D7S8	Gc
A	0.415	0.532	0.426	0.585	0.277
B	0.585	0.468	0.548	0.415	0.223
C	NA	NA	0.027	NA	0.500

PM loci observed allele frequencies in 96 unrelated Southwestern Hispanics					
Allele	LDLR	GYPA	HBGG	D7S8	Gc
A	0.563	0.656	0.344	0.682	0.271
B	0.438	0.344	0.609	0.318	0.208
C	NA	NA	0.047	NA	0.521

PM loci observed allele frequencies in 105 unrelated Chinese					
Allele	LDLR	GYPA	HBGG	D7S8	Gc
A	0.233	0.605	0.243	0.590	0.281
B	0.767	0.395	0.757	0.410	0.438
C	NA*	NA*	0.000	NA*	0.281

*NA = there is no C allele on the typing strips with the AmpliType PM PCR Amplification and Typing Kit (PE Biosystems, Foster City, CA) for LDLR, GYPA, and D7S8.

D1S80 ALLELE FREQUENCIES

Allele	African American (N=606)[a]	Caucasian (N=718)	Southeastern Hispanic (N=247)	Southwestern Hispanic (N=162)	Oriental (N=204)
15	0.000	0.000	0.000	0.003	0.000
16	0.002	0.001	0.004	0.019	0.034
17	0.028	0.002	0.012	0.003	0.025
18	0.073	0.237	0.225	0.222	0.152
19	0.003	0.003	0.004	0.006	0.022
20	0.032	0.018	0.010	0.019	0.007
21	0.115	0.021	0.030	0.025	0.034
22	0.081	0.038	0.028	0.019	0.017
23	0.014	0.012	0.014	0.000	0.017
24	0.234	0.378	0.316	0.315	0.230
25	0.045	0.046	0.059	0.093	0.027
26	0.006	0.020	0.008	0.006	0.000
27	0.008	0.007	0.012	0.022	0.047
28	0.130	0.063	0.081	0.074	0.076
29	0.053	0.052	0.079	0.019	0.042
30	0.009	0.008	0.018	0.071	0.123
31	0.054	0.072	0.051	0.056	0.093
32	0.007	0.006	0.006	0.003	0.012
33	0.004	0.003	0.004	0.003	0.005
34	0.086	0.001	0.008	0.003	0.005
35	0.002	0.003	0.000	0.000	0.005
36	0.001	0.004	0.014	0.006	0.005
37	0.000	0.001	0.006	0.000	0.007
38	0.000	0.000	0.000	0.000	0.000
39	0.003	0.003	0.002	0.006	0.005
40	0.000	0.000	0.000	0.000	0.000
41	0.002	0.000	0.000	0.006	0.007
>41[b]	0.007	0.001	0.008	0.003	0.002

[a]N = number of individuals.
[b]All alleles that migrate more slowly than the largest allele in the reference ladder (i.e. allele number 41) are placed in the >41 allele class.

D3S1358 Locus

	African American (n=210)	Caucasian (n=203)	SW Hispanic (n=209)	SE Hispanic (n=191)	Chinese (n=111)
<12	0.476	0.000	0.000	0.262	0.000
12	0.238	0.000	0.000	0.000	0.000
13	1.190	0.246	0.239	1.047	0.000
14	12.143	14.039	7.895	8.377	4.955
15	29.048	24.631	42.584	35.340	37.838
15.2	0.000	0.000	0.000	0.000	0.000
16	30.714	23.153	26.555	24.607	28.829
17	20.000	21.182	12.679	16.230	22.523
18	5.476	16.256	8.373	13.874	5.405
19	0.476	0.493	1.435	0.262	0.450
>19	0.238	0.000	0.239	0.000	0.000

n = number of individuals typed

vWA Locus

	African American (n=180)	Caucasian (n=196)	SW Hispanic (n=203)	SE Hispanic (n=240)	Chinese (n=111)
11	0.278	0.000	0.246	0.000	0.000
13	0.556	0.510	0.000	0.000	0.000
14	6.667	10.204	6.158	6.875	24.775
15	23.611	11.224	7.635	10.000	2.252
16	26.944	20.153	35.961	26.785	20.270
17	18.333	26.276	22.167	30.417	23.874
18	13.611	22.194	19.458	18.750	17.568
19	7.222	8.418	7.143	5.000	9.910
20	2.778	1.020	1.232	1.667	1.351
21	0.000	0.000	0.000	0.000	0.000

n = number of individuals typed

FGA Locus

	African American (n=180)	Caucasian (n=196)	SW Hispanic (n=203)	SE Hispanic (n=191)	Chinese (n=111)
<18	0.278	0.000	0.000	0.000	0.901
18	0.833	3.061	0.246	1.047	3.153
18.2	0.833	0.000	0.000	0.000	0.000

19	5.278	5.612	7.882	8.377	4.505
19.2	0.278	0.000	0.000	0.000	0.000
20	7.222	14.541	7.143	11.870	6.306
20.2	0.000	0.255	0.246	0.000	0.000
21	12.500	17.347	13.054	13.613	12.613
21.2	0.000	0.000	0.246	0.262	0.450
22	22.500	18.878	17.734	14.921	15.315
22.2	0.556	1.020	0.493	0.524	0.000
22.3	0.000	0.000	0.000	0.000	0.000
23	12.500	15.816	14.039	14.921	22.523
23.2	0.000	0.000	0.739	0.524	0.901
24	18.611	13.776	12.562	14.660	17.117
24.2	0.000	0.000	0.000	0.000	0.450
24.3	0.000	0.000	0.000	0.000	0.450
25	10.000	6.888	13.793	11.257	10.360
26	3.611	1.786	8.374	4.712	3.153
26.2	0.000	0.000	0.000	0.000	0.901
27	2.222	1.020	3.202	2.618	0.000
28	1.667	0.000	0.246	0.524	0.000
29	0.556	0.000	0.000	0.000	0.000
30	0.278	0.000	0.000	0.262	0.000
>30	0.278	0.000	0.000	0.000	0.000

n = number of individuals typed

D8S1179 Locus

	African American (n=180)	Caucasian (n=196)	SW Hispanic (n=203)	SE Hispanic (n=191)	Chinese (n=111)
8	0.278	1.786	0.246	1.571	0.000
9	0.556	1.020	0.246	1.047	0.000
10	2.500	10.204	9.360	9.686	7.658
11	3.611	5.867	6.158	5.236	10.811
12	10.833	14.541	12.069	10.733	16.216
13	22.222	33.929	32.512	35.602	17.117
14	33.333	20.153	24.631	21.204	19.820
15	21.389	10.969	11.576	12.042	18.468
16	4.444	1.276	2.463	2.618	7.658
17	0.833	0.255	0.739	0.262	1.351
18	0.000	0.000	0.000	0.000	0.450
19	0.000	0.000	0.000	0.000	0.450

n = number of individuals typed

D21S11 Locus

	African American (n=179)	Caucasian (n=196)	SW Hispanic (n=203)	SE Hispanic (n=191)	Chinese (n=111)
24.2	0.279	0.510	0.246	0.000	0.000
24.3	0.000	0.000	0.000	0.000	0.000
26	0.279	0.000	0.000	0.262	0.000
27	6.145	4.592	0.985	1.832	0.000
28	21.508	16.582	6.897	12.565	4.955
28.2	0.000	0.000	0.000	0.000	0.450
29	18.994	18.112	20.443	24.084	27.027
29.2	0.279	0.000	0.246	0.262	0.000
30	17.877	23.214	33.005	23.822	27.027
30.2	0.838	3.827	3.202	3.403	0.901
30.3	0.000	0.000	0.000	0.000	1.351
31	9.218	7.143	6.897	7.592	4.955
31.2	7.542	9.949	8.621	8.377	9.910
32	0.838	1.531	1.232	1.047	4.054
32.1	0.000	0.000	0.000	0.000	0.000
32.2	6.983	11.224	13.547	11.518	13.964
33	0.838	0.000	0.000	0.000	0.000
33.2	3.352	3.061	4.187	3.665	5.405
34	0.838	0.000	0.000	0.000	0.000
34.2	0.279	0.000	0.493	1.047	0.000
35	2.793	0.000	0.000	0.000	0.000
35.2	0.000	0.255	0.000	0.000	0.000
36	0.559	0.000	0.000	0.262	0.000
>36	0.559	0.000	0.000	0.262	0.000

n = number of individuals typed

D18S51 Locus

	African American (n=180)	Caucasian (n=196)	SW Hispanic (n=203)	SE Hispanic (n=191)	Chinese (n=111)
<11	0.093	1.276	0.493	0.785	0.901
11	0.556	1.276	1.232	1.571	0.000
12	5.833	12.755	10.591	13.613	1.802
13	5.556	12.245	16.995	11.780	15.766
13.2	0.556	0.000	0.000	0.000	0.000
14	6.389	17.347	16.995	13.089	22.973
14.2	0.000	0.000	0.000	0.000	0.000

15	16.667	12.755	13.793	19.110	18.018
15.2	0.000	0.000	0.000	0.000	0.000
16	18.889	10.714	11.576	14.136	10.811
17	16.389	15.561	13.793	10.733	10.360
18	13.056	9.184	5.172	5.497	4.054
19	7.778	3.571	3.695	4.712	4.505
20	5.556	2.551	1.724	2.880	4.054
21	1.111	0.510	1.970	1.047	3.153
21.2	0.000	0.000	0.000	0.000	0.000
22	0.556	0.255	0.739	0.785	1.802
>22	0.556	0.000	0.205	0.262	1.802

n = number of individuals typed

D5S818 Locus

	African American (n=180)	Caucasian (n=195)	SW Hispanic (n=203)	SE Hispanic (n=240)	Chinese (n=111)
7	0.278	0.000	6.158	2.292	2.703
8	5.000	0.000	0.246	1.042	0.000
9	1.389	3.077	5.419	5.000	9.459
10	6.389	4.872	6.650	4.583	15.315
11	26.111	41.026	42.118	39.375	34.685
12	35.556	35.385	29.064	31.667	24.775
13	24.444	14.615	9.606	15.417	10.811
14	0.556	0.769	0.493	0.625	1.351
15	0.000	0.256	0.246	0.000	0.901
>15	0.278	0.000	0.000	0.000	0.000

n = number of individuals typed

D13S317 Locus

	African American (n=179)	Caucasian (n=196)	SW Hispanic (n=203)	SE Hispanic (n=240)	Chinese (n=111)
7	0.000	0.000	0.000	0.000	0.450
8	3.631	9.949	6.650	11.458	23.423
9	2.793	7.653	21.921	11.458	16.216
10	5.028	5.102	10.099	7.708	15.766
11	23.743	31.888	20.197	30.625	24.775
12	48.324	30.867	21.675	22.917	13.964
13	12.570	10.969	13.793	10.833	4.505

14	3.631	3.571	5.665	5.000	0.901
15	0.279	0.000	0.000	0.000	0.000

n = number of individuals typed

D7S820 Locus

	African American (n=210)	Caucasian (n=203)	SW Hispanic (n=209)	SE Hispanic (n=240)	Chinese (n=111)
6	0.000	0.246	0.239	0.208	0.000
7	0.714	1.724	2.153	1.042	0.450
8	17.381	16.256	9.809	14.167	11.261
9	15.714	14.778	4.785	12.500	5.405
9.1	0.000	0.000	0.000	0.000	0.450
10	32.381	29.064	30.622	26.667	15.315
10.1	0.000	0.000	0.000	0.000	0.000
11	22.381	20.197	28.947	22.708	33.784
11.3	0.000	0.000	0.000	0.000	0.000
12	9.048	14.039	19.139	18.750	26.577
13	1.905	2.956	3.828	3.542	6.757
14	0.476	0.739	0.478	0.417	0.000

n = number of individuals typed

CSF1PO Locus

	African American (n=210)	Caucasian (n=203)	SW Hispanic (n=209)	SE Hispanic (n=240)	Chinese (n=111)
6	0.000	0.000	0.000	0.000	0.000
7	4.286	0.246	0.239	0.208	0.450
8	8.571	0.493	0.000	0.417	0.000
9	3.333	1.970	0.718	1.250	5.405
10	27.143	25.369	25.359	25.417	21.622
10.3	0.000	0.246	0.000	0.000	0.000
11	20.476	30.049	26.555	29.583	27.477
12	30.000	32.512	39.234	35.625	36.486
13	5.476	7.143	6.459	6.875	8.108
14	0.714	1.478	0.957	0.417	0.450
15	0.000	0.493	0.478	0.208	0.000

n = number of individuals typed

TPOX Locus

	African American (n=209)	Caucasian (n=203)	SW Hispanic (n=209)	SE Hispanic (n=240)	Chinese (n=111)
6	8.612	0.000	0.478	0.417	0.000
7	2.153	0.246	0.478	0.208	0.000
8	36.842	54.433	55.502	50.625	53.153
9	18.182	12.315	3.349	8.333	12.613
10	9.330	3.695	3.349	6.250	4.054
11	22.488	25.369	27.273	27.708	27.477
12	2.392	3.941	9.330	6.458	2.703
13	0.000	0.000	0.239	0.000	0.000

n = number of individuals typed

THO1 Locus

	African American (n=210)	Caucasian (n=203)	SW Hispanic (n=209)	SE Hispanic (n=240)	Chinese (n=111)
5	0.000	0.000	0.239	0.000	0.000
6	10.952	22.660	23.206	21.250	9.009
7	44.048	17.241	33.732	25.208	29.279
8	18.571	12.562	8.134	10.417	6.306
8.3	0.000	0.246	0.000	0.000	0.000
9	14.524	16.502	10.287	18.542	46.396
9.3	10.476	30.542	24.163	23.542	4.505
10	1.429	0.246	0.239	1.042	4.505

n = number of individuals typed

D16S539 Locus

	African American (n=209)	Caucasian (n=202)	SW Hispanic (n=208)	SE Hispanic (n=240)	Chinese (n=111)
8	3.589	1.980	1.683	2.292	0.000
9	19.856	10.396	7.933	14.583	27.027
10	11.005	6.683	17.308	9.583	10.811
11	29.426	27.228	31.490	28.125	30.631
12	18.660	33.911	28.606	25.417	23.874
13	16.507	16.337	10.337	16.250	4.054
14	0.957	3.218	2.404	3.542	3.604
15	0.000	0.248	0.240	0.208	0.000

n = number of individuals typed

Reference sequence for mitochondrial DNA regions HV1 and HV2

HV1

16024 T	16025 T	16026 C	16027 T	16028 T	16029 T	16030 C
16031 A	16032 T	16033 G	16034 G	16035 G	16036 G	16037 A
16038 A	16039 G	16040 C	16041 A	16042 G	16043 A	16044 T
16045 T	16046 T	16047 G	16048 G	16049 G	16050 T	16051 A
16052 C	16053 C	16054 A	16055 C	16056 C	16057 C	16058 A
16059 A	16060 G	16061 T	16062 A	16063 T	16064 T	16065 G
16066 A	16067 C	16068 T	16069 C	16070 A	16071 C	16072 C
16073 C	16074 A	16075 T	16076 C	16077 A	16078 A	16079 C
16080 A	16081 A	16082 C	16083 C	16084 G	16085 C	16086 T
16087 A	16088 T	16089 G	16090 T	16091 A	16092 T	16093 T
16094 T	16095 C	16096 G	16097 T	16098 A	16099 C	16100 A
16101 T	16102 T	16103 A	16104 C	16105 T	16106 G	16107 C
16108 C	16109 A	16110 G	16111 C	16112 C	16113 A	16114 C
16115 C	16116 A	16117 T	16118 G	16119 A	16120 A	16121 T
16122 A	16123 T	16124 T	16125 G	16126 T	16127 A	16128 C
16129 G	16130 G	16131 T	16132 A	16133 C	16134 C	16135 A
16136 T	16137 A	16138 A	16139 A	16140 T	16141 A	16142 C
16143 T	16144 T	16145 G	16146 A	16147 C	16148 C	16149 A
16150 C	16151 C	16152 T	16153 G	16154 T	16155 A	16156 G
16157 T	16158 A	16159 C	16160 A	16161 T	16162 A	16163 A
16164 A	16165 A	16166 A	16167 C	16168 C	16169 C	16170 A
16171 A	16172 T	16173 C	16174 C	16175 A	16176 C	16177 A
16178 T	16179 C	16180 A	16181 A	16182 A	16183 A	16184 C
16185 C	16186 C	16187 C	16188 C	16189 T	16190 C	16191 C
16192 C	16193 C	16194 A	16195 T	16196 G	16197 C	16198 T
16199 T	16200 A	16201 C	16202 A	16203 A	16204 G	16205 C
16206 A	16207 A	16208 G	16209 T	16210 A	16211 C	16212 A
16213 G	16214 C	16215 A	16216 A	16217 T	16218 C	16219 A
16220 A	16221 C	16222 C	16223 C	16224 T	16225 C	16226 A
16227 A	16228 C	16229 T	16230 A	16231 T	16232 C	16233 A
16234 C	16235 A	16236 C	16237 A	16238 T	16239 C	16240 A
16241 A	16242 C	16243 T	16244 G	16245 C	16246 A	16247 A
16248 C	16249 T	16250 C	16251 C	16252 A	16253 A	16254 A
16255 G	16256 C	16257 C	16258 A	16259 C	16260 C	16261 C
16262 C	16263 T	16264 C	16265 A	16266 C	16267 C	16268 C
16269 A	16270 C	16271 T	16272 A	16273 G	16274 G	16275 A
16276 T	16277 A	16278 C	16279 C	16280 A	16281 A	16282 C
16283 A	16284 A	16285 A	16286 C	16287 C	16288 T	16289 A
16290 C	16291 C	16292 C	16293 A	16294 C	16295 C	16296 C
16297 T	16298 T	16299 A	16300 A	16301 C	16302 A	16303 G
16304 T	16305 A	16306 C	16307 A	16308 T	16309 A	16310 G
16311 T	16312 A	16313 C	16314 A	16315 T	16316 A	16317 A
16318 A	16319 G	16320 C	16321 C	16322 A	16323 T	16324 T
16325 T	16326 A	16327 C	16328 C	16329 G	16330 T	16331 A
16332 C	16333 A	16334 T	16335 A	16336 G	16337 C	16338 A
16339 C	16340 A	16341 T	16342 T	16343 A	16344 C	16345 A
16346 G	16347 T	16348 C	16349 A	16350 A	16351 A	16352 T
16353 C	16354 C	16355 C	16356 T	16357 T	16358 C	16359 T
16360 C	16361 G	16362 T	16363 C	16364 C	16365 C	16366 C
16367 A	16368 T	16369 G	16370 G	16371 A	16372 T	16373 G

16374 A	16375 C	16376 C	16377 C	16378 C	16379 C	16380 C	
16381 T	16382 C	16383 A	16384 G	16385 A	16386 T	16387 A	
16388 G	16389 G	16390 G	16391 G	16392 T	16393 C	16394 C	
16395 C	16396 T	16397 T	16398 G	16399 A	16400 C		

HV2

19 C	20 T	21 A	22 T	23 T	24 A	25 A	26 C	27 C
28 A	29 C	30 T	31 C	32 A	33 C	34 G	35 G	36 G
37 A	38 G	39 C	40 T	41 C	42 T	43 C	44 C	45 A
46 T	47 G	48 C	49 A	50 T	51 T	52 T	53 G	54 G
55 T	56 A	57 T	58 T	59 T	60 T	61 C	62 G	63 T
64 C	65 T	66 G	67 G	68 G	69 G	70 G	71 G	72 T
73 A	74 T	75 G	76 C	77 A	78 C	79 G	80 C	81 G
82 A	83 T	84 A	85 G	86 C	87 A	88 T	89 T	90 G
91 C	92 G	93 A	94 G	95 A	96 C	97 G	98 C	99 T
100 G	101 G	102 A	103 G	104 C	105 C	106 G	107 G	108 A
109 G	110 C	111 A	112 C	113 C	114 C	115 T	116 A	117 T
118 G	119 T	120 C	121 G	122 C	123 A	124 G	125 T	126 A
127 T	128 C	129 T	130 G	131 T	132 C	133 T	134 T	135 T
136 G	137 A	138 T	139 T	140 C	141 C	142 T	143 G	144 C
145 C	146 T	147 C	148 A	149 T	150 C	151 C	152 T	153 A
154 T	155 T	156 A	157 T	158 T	159 T	160 A	161 T	162 C
163 G	164 C	165 A	166 C	167 T	168 T	169 A	170 C	171 G
172 T	173 T	174 T	175 A	176 A	177 T	178 A	179 T	180 T
181 A	182 C	183 A	184 G	185 G	186 C	187 G	188 A	189 A
190 C	191 A	192 T	193 A	194 C	195 T	196 T	197 A	198 C
199 T	200 A	201 A	202 A	203 G	204 T	205 G	206 T	207 G
208 T	209 T	210 A	211 A	212 T	213 T	214 A	215 A	216 T
217 T	218 A	219 A	220 T	221 G	222 C	223 T	224 T	225 G
226 T	227 A	228 G	229 G	230 A	231 C	232 A	233 T	234 A
235 A	236 T	237 A	238 A	239 T	240 A	241 A	242 C	243 A
244 A	245 T	246 T	247 G	248 A	249 A	250 T	251 G	252 T
253 C	254 T	255 G	256 C	257 A	258 C	259 A	260 G	261 C
262 C	263 A	264 C	265 T	266 T	267 T	268 C	269 C	270 A
271 C	272 A	273 C	274 A	275 G	276 A	277 C	278 A	279 T
280 C	281 A	282 T	283 A	284 A	285 C	286 A	287 A	288 A
289 A	290 A	291 A	292 T	293 T	294 T	295 C	296 C	297 A
298 C	299 C	300 A	301 A	302 A	303 C	304 C	305 C	306 C
307 C	308 C	309 C	310 T	311 C	312 C	313 C	314 C	315 C
316 G	317 C	318 T	319 T	320 C	321 T	322 G	323 G	324 C
325 C	326 A	327 C	328 A	329 G	330 C	331 A	332 C	333 T
334 T	335 A	336 A	337 A	338 C	339 A	340 C	341 A	342 T
343 C	344 T	345 C	346 T	347 G	348 C	349 C	350 A	351 A
352 A	353 C	354 C	355 C	356 C	357 A	358 A	359 A	360 A
361 A	362 C	363 A	364 A	365 A	366 G	367 A	368 A	369 C
370 C	371 C	372 T	373 A	374 A	375 C	376 A	377 C	378 C
379 A	380 G	381 C	382 C	383 T	384 A	385 A	386 C	387 C
388 A	389 G	390 A	391 T	392 T	393 T	394 C	395 A	396 A
397 A	398 T	399 T	400 T	401 T	402 A	403 T	404 C	405 T
406 T	407 T	408 T	409 G	410 G				

REFERENCES

1. Alkhayat, A., F. Alshamali, and B. Budowle. 1996. Population data on the PCR-based loci, LDLR, GYPA, HBGG, D7S8, Gc, HLA-DQA1, and D1S80 from Arabs from Dubai. Forensic Sci. Int. *81*:29-34.
2. Alonso A., P. Martin, C. Albarran, and M. Sancho. 1993. Amplified fragment-length polymorphism analysis of the VNTR locus D1S80 in central Spain. Int. J. Legal Med. *106*:311-314.
3. Ambach, E., W. Parson, H. Niederstatter, and B. Budowle. 1997. Austrian Caucasian population data for the quadruplex plus amelogenin: refined mutation rate for HumvWFA31/A. J. Forensic Sci. *42*:1136-1139.
4. Baird, M., I. Balazs, A. Giusti, G. Miyasaki, L. Nicholas, K. Wexler, E. Kanter, J. Glassberg, F. Allen, P. Rubinstein, and L. Sussman. 1986. Allele frequency distribution of two highly polymorphic DNA sequences in three ethnic groups and its application to the determination of paternity. Amer. J. Hum. Genet. *39*:489-501.
5. Balamurugan, K., H. Abdel-Rehman, G.T. Duncan, B. Budowle, S. Anderson, J. Macechko, and M. Tahir. 1998. Distribution of D1S80 alleles in the Jordanian population. Int. J. Legal Med. *111*:276-277.
6. Balazs, I., M. Baird, M. Clyne, and E. Meade. 1989. Human population genetic studies of five hypervariable DNA loci. Amer. J. Hum. Genet. *44*:182-190.
7. Bayoumi, R.A., L.I. Al-Gazali, U. Jaffer, M.S.A. Nur-E-Kamal, A. Dawodu, A. Bener, V. Eapen, and B. Budowle. 1997. United Arab Emirate population data on three short tandem repeat loci: HUMTHO1, TPOX, and CSF1PO -derived using multiplex PCR and manual typing. Electrophoresis *18*:1637-1640.
8. Bell, B., B. Budowle, B. Martinez-Jarreta, Y. Casalod, E. Abecia, and M. Castellano. 1997. Distribution types for six PCR-based loci LDLR, GYPA, HBGG, D7S8, Gc, and HLA-DQA1 in central Pyrenees and Teruel (Spain). J. Forensic Sci. *42*:510-513.
9. Budowle, B. 1995. The effects of inbreeding on DNA profile frequency estimates using PCR-based loci. Genetica *96*:21-25.
10. Budowle, B. 1995. Finns and Italians as categorical support for population substructure, the effects of inbreeding, and the probability of innocence -are these real concerns regarding DNA forensic statistics? p. 43-48. *In* Fifth International Symposium on Human Identification 1994, Promega Corporation, Madison, WI.
11. Budowle, B., F.S. Baechtel, and R. Fejeran. 1998. Polymarker, HLA-DQA1, and D1S80 allele frequency data in Chamorro and Filipino populations from Guam. J. Forensic Sci. *43*:1195-1198.
12. Budowle, B., F.S. Baechtel, J.B. Smerick, K.W. Presley, A.M. Giusti, G. Parsons, M. Alevy, and R. Chakraborty. 1995. D1S80 population data in African Americans, Caucasians, Southeastern Hispanics, Southwestern Hispanics, and Orientals. J. Forensic Sci. *40*:38-44.
13. Budowle, B. and A. Giusti. 1995. Fixed Bin Frequency Distributions for the VNTR Locus D5S110 in General United States Reference Databases. J. Forensic Sci. *40*:94-96.
14. Budowle, B., A. Giusti, J. Waye, F.S. Baechtel, R. Fourney, D. Adams, L. Presley, H. Deadman, and K. Monson. 1991. Fixed-bin analysis for statistical evaluation of continuous distributions of allelic data from VNTR loci, for use in forensic comparisons. Amer. J. Hum. Genet. *48*:841-855.
15. Budowle, B., L.B. Jankowski, H.W. Corey, N.T. Swec, S. Freck-Tootell, J.A. Pino, R. Schwartz, C.A. Kelley, and M.L. Tarver. 1997. Evaluation of independence assumptions for PCR-based and protein-based genetic markers in New Jersey Caucasians. J. Forensic Sci. *42*:223-225.
16. Budowle, B., B.W. Koons, and T.R. Moretti. 1998. Subtyping of the HLA-DQA1 locus and independence testing with PM and STR/VNTR loci. J. Forensic Sci. *43*:657-660.
17. Budowle, B., J.A. Lindsey, J.A. DeCou, B.W. Koons, A.M. Giusti, and C.T. Comey. 1995. Validation and population studies of the loci LDLR, GYPA, HBGG, D7S8, and Gc (PM loci), and HLA-DQα using a multiplex amplification and typing procedure. J. Forensic Sci. *40*:45-54.
18. Budowle, B. and K.L. Monson. 1998. Database size for frequency estimation of PCR profiles, p. 26-37. *In* Eighth International Symposium on Human Identification 1997, Promega Corporation, Madison, WI.
19. Budowle, B., K. Monson, K. Anoe, F.S. Baechtel, D. Bergman, E. Buel, P. Campbell, M. Clement, H. Corey, L. Davis, A. Dixon, et al. 1991. A preliminary report on binned general population data on six VNTR loci in Caucasians, Blacks and Hispanics from the United States. Crime Lab. Digest *18*:9-26.
20. Budowle, B., K.L. Monson, and R. Chakraborty. 1996. Estimating minimum allele frequencies

for DNA profile frequency estimates for PCR-based loci. Int. J. Legal Med. *108*:173-176.

21. Budowle, B., K. Monson, A. Giusti, and B. Brown. 1994. The assessment of frequency estimates of *Hae*III-generated VNTR profiles in various reference databases. J. Forensic Sci. *39*:319-352.

22. Budowle, B., K. Monson, A. Giusti, and B. Brown. 1994. Evaluation of *Hinf*I-generated VNTR profile frequencies determined using various ethnic databases. J. Forensic Sci. *39*:988-1008.

23. Budowle, B., K. Monson, and A. Giusti. 1994. A reassessment of frequency estimates of *Pvu*II-generated VNTR profiles in a Finnish, an Italian, and a General United States Caucasian database: no evidence for ethnic subgroups affecting forensic estimates. Amer. J. Hum. Genet. *55*:533-539.

24. Budowle, B. and T.R. Moretti. 1999. Genotype profiles for six population groups at the 13 CODIS short tandem repeat core loci and other PCR-based loci. Forensic Science Communications 1 July 1999. Available: http://www.fbi.gov/programs/lab/fsc/current/budowle.htm.

25. Budowle, B., T.R. Moretti, A.L. Baumstark, D.A. Defenbaugh, and K.M. Keys. 1999. Population data on the thirteen CODIS core short tandem repeat loci in African Americans, U.S. Caucasians, Hispanics, Bahamians, Jamaicans, and Trinidadians. J. Forensic Sci. *44*:1277-1286.

26. Budowle, B., L.T. Nhari, T.R. Moretti, S.B. Kanoyangwa, E. Masuka, D.A. Defenbaugh, and J.B. Smerick. 1997. Zimbabwe black population data on the six short tandem repeat loci - CSF1PO, TPOX, THO1, D3S1358, VWA, and FGA. Forensic Sci. Int. *90*:215-221.

27. Budowle, B., M.R. Wilson, J.A. DiZinno, C. Stauffer, M.A. Fasano, M.M. Holland, and K.L. Monson. 1999. Mitochondrial DNA regions HVI and HVII population data. Forensic Sci. Int. *103*:23-35.

28. Cariolou, M.A., P. Manoli, M. Christophorou, E. Bashiardes, A. Karagrigoriou, and B. Budowle. 1998. Greek Cypriot allele and genotype frequencies of Amplitype PM-DQA1 and D1S80 loci. J. Forensic Sci. *43*:661-664.

29. Chakraborty, R., M. Fornage, R. Guegue, and E. Boerwinkle. 1991. Population genetics of hypervariable loci: analysis of PCR based VNTR polymorphism within a population. *In* T. Burke, G. Dolf, A. Jeffreys, and R. Wolff (Eds.), DNA Fingerprinting: Approaches and Applications, Birkhäuser, Inc., Basel, Boston, Berlin.

30. Chakraborty, R. and K. Kidd. 1991. The utility of DNA typing in forensic work. Science *254*:1735-1739.

31. Chakraborty, R. 1992. Effects of population subdivision and allele frequency differences on interpretation of DNA typing data for human identification, p. 205-222. *In* Proceedings from the Third International Symposium on Human Identification, Promega Corporation, Madison, WI.

32. Chakraborty, R., M. de Andrade, S. Daiger, and B. Budowle. 1992. Apparent heterozygote deficiencies observed in DNA typing data and their implications in forensic applications. Ann. Hum. Genet. *56*:45-57.

33. Chakraborty, R. and L. Jin. 1992. Heterozygote deficiency, population substructure and their implications in DNA fingerprinting. Hum. Genet. *88*:267-272.

34. Chakraborty, R., D.N. Stivers, B. Su, Y. Zhong, and B. Budowle. 1999. The utility of STR loci beyond human identification: Implications for the development of new DNA typing systems. Electrophoresis *20*:1682-1686.

35. Chakraborty, R., Y. Zhong, L. Jin, and B. Budowle. 1994. Nondetectability of restriction fragments and independence of DNA fragment sizes within and between loci in RFLP typing of DNA. Amer. J. Hum. Genet. *55*:391-401.

36. Comey, C.T. and B. Budowle. 1991. Validation studies on the analysis of the HLA-DQ alpha locus using the polymerase chain reaction. J. Forensic Sci. *36*:1633-1648.

37. Crouse, C.A., W.J. Feuer, D.C. Nippes, S.C. Hutto, K.S. Barnes, D. Coffman, S.H. Livingston, L. Ginsberg, and D.E. Glidewell. 1994. Analysis of HLA DQ alpha gene and genotype frequencies in populations from Florida. J. Forensic Sci. *39*:731-742.

38. Deka, R., R. Chakraborty, and R. Ferrell. 1991. A population genetic study of six VNTR loci in three ethnically defined populations. Genomics *11*:83-92.

39. DeStefano, F., L. Casarino, A. Mannucci, L. Delfino, M. Canale, and G.B. Ferrara. 1992. HLA DQA1 allele and genotype frequencies in a Northern Italian population. Forensic Sci. Int. *55*:59-66.

40. Devlin, B., N. Risch, and K. Roeder. 1990. No excess of homozygosity at loci used for DNA fingerprinting. Science *249*:1416-1420.

41. Devlin, B., N. Risch, and K. Roeder. 1991. Estimation of allele frequencies for VNTR loci. Amer. J. Hum. Genet. *48*:662-676.

42. Devlin, B. and N. Risch. 1992. A note on Hardy-Weinberg equilibrium of VNTR data by using the Federal Bureau of Investigation's fixed-bin method. Amer. J. Hum. Genet. *51*:549-553.

43.Devlin, B. and N. Risch. 1992. Ethnic differentiation at VNTR loci, with special reference to forensic applications. Amer. J. Hum. Genet. *51*:534-548.

44.Devlin, B. and N. Risch. 1993. Physical properties of VNTR data, and their impact on a test of allelic independence. Amer. J. Hum. Genet. *53*:324-329.

45.Devlin, B., N. Risch, and K. Roeder. 1994. Comments on the statistical aspects of the NRC's Report on DNA typing. J. Forensic Sci. *39*:28-40.

46.Drobnic, K. and B. Budowle. 2000. The analysis of three Short Tandem Repeat (STR) loci in the Slovene population by multiplex PCR. J. Forensic Sci. *45*:893-895.

47.Drobnic, K., A. Regent, and B. Budowle. 2000. The Slovenian population data on the PCR based HLA-DQA1, LDLR, GYPA, HBGG, D7S8, GC, and D1S80. J. Forensic Sci. *45*:689-691.

48.Edwards, A., A. Civitello, H.A. Hammond, and C.T. Caskey. 1991. DNA typing and genetic mapping with trimeric and tetrameric tandem repeats. Amer. J. Hum. Genet. *49*:746-756.

49.Edwards, A., H.A. Hammond, L. Jin, C.T. Caskey, and R. Chakraborty. 1992. Genetic variation at five trimeric and tetrameric repeat loci in four human population groups. Genomics *12*:241-253.

50.Entrala, C., M. Lorente, J.A. Lorente, J.C. Alvarez, T. Moretti, B. Budowle, and E. Villanueva. 1998. Fluorescent multiplex analysis of nine STR loci and the amelogenin locus: Spanish population data. Forensic Sci. Int. *98*:179-183.

51.Entrala, C., J.A. Lorente, M. Lorente, J.C. Alvarez, B. Budowle, and E. Villanueva. 1999. Spanish population data on the loci D13S317, D7S820, D16S539 generated using silver staining (SilverSTR III™ Multiplex). J. Forensic Sci. *44*:1032-1034.

52.Furedi, S., B. Budowle, J. Woller, and Z. Padar. 1996. Hungarian population data on 6 STR loci -HUMVWA31/1, HUMTHO1, HUMCSF1PO, HUMFES/FPS, HUMTPOX, AND HUMH-PRTB - derived using multiplex PCR amplification and manual typing. Int. J. Legal Med. *109*:100-101.

53.Garcia, O., P. Martin, B. Budowle, J. Uriarte, C. Albarran, and A. Alonso. 1998. Basque Country autochthonous population data on 7 short tandem repeat loci. Int. J. Legal Med. *111*:162-164.

54.Garofano, L., M. Pizzamiglio, C. Vecchio, G. Lago, T Floris, G. D'Errico, G. Brembilla, A. Romano, and B. Budowle. 1998. Italian population data on thirteen short tandem repeat loci: THO1, D21S11, D18S51, VWA, FGA, D8S1179, TPOX, CSF1PO, D16S539, D7S820, D13S317, D5S818, D3S1358. Forensic Sci. Int. *97*:53-60.

55.Gehrig, C., M. Hochmeister, U.V. Borer, and B. Budowle. 1999. Swiss Caucasian population DNA data for 13 STR loci using AmpFlSTR Profiler Plus and Cofiler PCR amplification kits. J. Forensic Sci. *44*:1035-1038.

56.Gehrig, C., H. Hochmeister, B. Budowle, and R. Reynolds. 1996. Subtyping the HLA-DQA1 locus in the Swiss Population. Forensic Sci. Int. *83*:27-30.

57.Hayes, J.M., B. Budowle, and M. Freund. 1995. Arab population data on the PCR-based loci: HLA-DQA1, LDLR, GYPA, HBGG, D7S8, Gc, and D1S80. J. Forensic Sci. *40*:888-892.

58.Helmuth, R., N. Fildes, E. Blake, M.C. Luce, J. Chimera, R. Madej, C. Gorodezky, M. Stoneking, N. Schmill, W. Klitz, R. Higuchi, and H.A. Erlich. 1990. HLA-DQ alpha allele and genotype frequencies in various human populations, determined by using enzymatic amplification and oligonucleotide probes. Amer. J. Hum. Genet. *47*:515-523.

59.Herrin, G. 1993. Probability of matching RFLP patterns from unrelated individuals. Amer. J. Hum. Genet. *52*:491-497.

60.Herrin, G. 1992. A comparison of models used for calculation of RFLP pattern frequencies. J. Forensic Sci. *37*:1640-1651.

61.Hochmeister, M.N., B. Budowle, U.V. Borer, and R. Dirnhofer. 1994. Swiss population data on the loci HLA-DQα, LDLR, GYPA, HBGG, D7S8, Gc, and D1S80. Forensic Sci. Int. *67*:175-184.

62.Huang, N.E. and B. Budowle. 1995. Fixed bin population data for the VNTR loci D1S7, D2S44, D4S139, D5S110, and D17S79 in Chinese from Taiwan. J. Forensic Sci. *40*:287-290.

63.Huang, N.E. and B. Budowle. 1995. Chinese population data on the PCR-based loci HLA-DQα, LDLR, GYPA, HBGG, D7S8, and Gc. Hum. Hered. *45*:34-40.

64.Huang, N.E., R. Chakraborty, and B. Budowle. 1994. D1S80 allele frequencies in a Chinese population. Int. J. Legal Med. *107*:118-120.

65.Huang, N.E., J. Schumm, and B. Budowle. 1995. Chinese Population Data on Three Tetrameric Short Tandem Repeat Loci - HUMTHO1, TPOX, AND CSF1PO - Derived Using Multiplex PCR and Manual Typing. Forensic Sci. Int. *71*:131-136.

66.Jorquera, H. and B. Budowle. 1998. Chilean population data on ten PCR-based loci. J. Forensic Sci. *43*:171-173.

67. Jankowski, L.B., B. Budowle, N.T. Swec, J.A. Pino, S. Freck-Tootell, H.W. Corey, R. Schwartz, E.J. LaRue, W.L. Rochin, C.J. Kearner, and M.L. Tarver. 1998. New Jersey Caucasian, African American, and Hispanic population data on the PCR-based loci HLA-DQA1, LDLR, GYPA, HBGG, D7S8, and Gc. J. Forensic Sci. *43*:1037-1040.

68. Keys, K.M., B. Budowle, S. Andelinovic, M. Definis-Gojanovic, I. Drmic, M. Mladen, and D. Primorac. 1996. Northern and southern Croatian population data on seven PCR-based loci. Forensic Sci. Int. *81*:191-199.

69. Kloosterman, A.D., P. Daselaar, B. Budowle, and E.L. Riley. 1992. Population genetic study on the HLA DQα and the D1S80 locus in Dutch Caucasians, p. 329-344. *In* Proceedings from the Third International Symposium on Human Identification 1992, Promega Corporation, Madison, WI.

70. Kloosterman, A.D., B. Budowle, and P. Daselaar. 1993. PCR-amplification and detection of the human D1S80 VNTR locus: Amplification conditions, population genetics, and application in forensic analysis. Int. J. Legal Med. *105*:257-264.

71. Kloosterman, A.D., B. Budowle, and E.L. Riley. 1993. Population data of the HLA DQα locus in Dutch Caucasians. Comparison with seven other population studies. Int. J. Legal Med. *105*: 233-238.

72. Koh, C-L. and D.G. Benjamin. 1994. HLA-DQa genotype and allele frequencies in Malays, Chinese, and Indians in the Malaysian population. Hum. Hered. *44*:150-155.

73. Kupferschmid, T.D., T. Calicchio, and B. Budowle. 1999. Maine Caucasian population DNA database using twelve short tandem repeat loci. J. Forensic Sci. *44*:392-395.

74. Lander, E. 1989. DNA fingerprinting on trial. Nature *339*:501-505.

75. Lander, E.S. and B. Budowle. 1994. DNA fingerprinting controversy laid to rest. Nature *371*:735-738.

76. Lareu, M.V., I. Munoz, M.S. Rodriguez, C. Vide, and A. Carracedo. 1993. The distribution of HLA DQA1 and D1S80 (PMCT118) alleles and genotypes in the population of Galicia and Central Portugal. Int. J. Legal Med. *106*:124-128.

77. Latter, B. 1980. Genetic differences within and between populations of the major human subgroups. Amer. Natural. *116*:220-237.

78. Lewontin, R. and D. Hartl. 1991. Population genetics in forensic DNA typing. Science *254*:1745-1750.

79. Li, C. and A. Chakravarti. 1994. DNA profile similarity in a subdivided population. Hum. Hered. *44*:100-109.

80. Lorente, M., J.A. Lorente, M.R. Wilson, B. Budowle, and E. Villanueva. 1997. Spanish population data on seven loci: D1S80, D17S5, HUMTHO1, HUMVWA, ACTBP2, D21S11, and DQA1. Forensic Sci. Int. *86*:163-171.

81. Martin, P., A. Alonso, B. Budowle, C. Albarran, O. Garcia, and M. Sancho. 1995. Spanish population data on seven tetrameric short tandem repeat loci. Int. J. Legal Med. *108*:145-149.

82. Martinez-Jarreta, B., P. Diaz Roche, B. Budowle, E. Abecia, M. Castellano, and Y. Casalod. 1998. Pyrenean population data on 3 tetrameric short tandem repeat loci—HUMTHO1, TPOX, and CSF1PO—derived using a STR multiplex system, p. 312-314. *In* B. Olaisen, B. Brinkmann, and P.J. Lincoln (Eds.), Advances in Forensic Haemogenetics 7, Elsevier, Amsterdam.

83. Martinez-Jarreta, B., B. Budowle, E. Abecia, B. Bell, Y. Casalod, and M. Castellano. 1998. PM and D1S80 in the Zaragoza population of North Spain. J. Forensic Sci. *43*:1094-1096.

84. Monson, K. and B. Budowle. 1993. A comparison of the fixed bin method with the floating bin and direct count methods: effect of VNTR profile frequency estimation and reference population. J. Forensic Sci. *38*:1037-1050.

85. Monson, K.L., J.P. Moisan, O. Pascal, M. McSween, D. Aubert, A. Giusti, B. Budowle, and L. Lavergne. 1995. Description and analysis of allele distribution for four VNTR markers in French and French Canadian populations. Hum. Hered. *45*:135-143.

86. Morton, N. 1992. Genetic structure of forensic populations. Proc. Natl. Acad. Sci. *89*:2556-2560.

87. Moura-Neto, R.S. and B. Budowle. 1997. Fixed bin population data for the VNTR loci D1S7, D2S44, D4S139, D5S110, D10S28, and D14S13 in a population sample from Rio de Janeiro, Brazil. J. Forensic Sci. *42*:926-928.

88. Nagai, A., S. Yamada, Y. Bunai, and I. Ohya. 1994. Analysis of the VNTR locus D1S80 in a Japanese population. Int. J. Legal Med. *106*:268-270.

89. National Research Council. 1992. DNA Technology in Forensic Science, National Academy Press, Washington, DC.

90. National Research Council. 1996. The Evaluation of Forensic DNA Evidence, National Academy Press, Washington, DC.

91. Padula, R.A., D. Gangitano, G.L. Juvenal, and B. Budowle. 1999. Allele frequencies in the population of Buenos Aires (Argentina) using AmpliType PM + DQA1. J. Forensic Sci. *44*:1320.

92. Peterson, B.L., B. Su, R. Chakraborty, B. Budowle, and R.E. Gaensslen. 2000. World population data for the HLA-DQA1, PM and D1S80 loci with least and most common profile frequencies for combinations of loci estimated following NRC II guidelines. J. Forensic Sci. *45*:118-146.

93. Risch, N. and B. Devlin. 1992. On the probability of matching DNA fingerprints. Science *255*:717-720.

94. Sajantila, A., B. Budowle, M. Strom, V. Johnsson, M. Lukka, L. Peltonen, and C. Ehnholm. 1992. PCR amplfication of alleles at the D1S80 locus: comparison of a Finnish and a North American Caucasian population sample, and forensic case-work evaluation. Amer. J. Hum. Genet. *50*:816-825.

95. Santos, S.M.M., B. Budowle, J.B. Smerick, K.M. Keys, B.W. Koons, and T.R. Moretti. 1996. Portuguese population data on the six short tandem repeat loci - CSF1PO, TPOX, THO1, D3S1358, VWA, and FGA. Forensic Sci. Int. *83*:229-235.

96. Scholl, S., B. Budowle, K. Radecki, and M. Salvo. 1996. Navajo, Pueblo, and Sioux population data on the loci HLA-DQA1, LDLR, GYPA, HBGG, D7S8, Gc, and D1S80. J. Forensic Sci. *41*:47-51.

97. Skowasch, K., P. Wiegand, and B. Brinkmann. 1992. pMCT118 (D1S80): a new allelic ladder and improved electrophoretic separation lead to the demonstration of 28 alleles. Int. J. Legal Med. *105*:165-168.

98. Sovinski, S.M., L.S. Baird, B. Budowle, J.F. Caruso, P.S. Davender, M.S. Cheema, G.T. Duncan, P.P. Hamby, A.S. Masibay, V.J. Sharma, and M.A. Tahir. 1996. The development of a deoxyribonucleic acid (DNA) restriction fragment length polymorphism (RFLP) database for Punjabis in East Punjab, India. Forensic Sci. Int. *79*:187-198.

99. Sugiyama, E., K. Honda, Y. Katsuyama, S. Uchiyama, A. Tsuchikane, M. Ota, and H. Fukushima. 1993. Allele frequency distribution of the D1S80 (pMCT118) locus polymorphism in the Japanese population by the polymerase chain reaction. Int J. Legal Med. *106*:111-114.

100. Sullivan, K.M., P. Gill, D. Lingard, and J.E. Lygo. 1992. Characterization of HLA-DQ alpha for forensic purposes. Allele and genotype frequencies in British Caucasian, Afro-Caribbean and Asian populations. Int. J. Legal Med. *105*:17-20.

101. Tahir, M.A., A.Q. Alkhayat, F.A. Shamali, B. Budowle, and C.E. Novick. 1997. Distribution of HLA-DQA1 alleles in Arab and Pakistani individuals from Dubai, United Arab Emirates. Forensic Sci. Int. *85*:219-223.

102. Tahir, M., J. Caruso, B. Budowle, and G.E. Novick. 1997. Distribution of HLA-DQA1, and Polymarker (LDLR, GYPA, HBGG, D7S8, and GC) alleles in Arab and Pakistani populations living in Abu Dhabi, United Arab Emirates. J. Forensic Sci. *42*:914-918.

103. Thymann, M., L.J. Nellemann, G. Masumba, L. Iregns-Moeller, and N. Morling. 1993. Analysis of the locus D1S80 by amplified fragment length polymorphism technique (AMP-FLP). Frequency distribution in Danes, intra and inter laboratory reproducibility of the technique. Forensic Sci. Int. *60*:47-56.

104. Tomscy, C.S., C.J. Basten, B. Budowle, B.A. Giles, S. Ermlick, and S. Gotwald. 1999. Use of combined frequencies for RFLP and PCR based loci in determining match probability. J. Forensic Sci. *44*:385-388.

105. VNTR Population Data: A Worldwide Study. Volumes I-IV. 1993. U.S. Government Printing Office, Washington, D.C.

106. Weir, B.S. 1992. Independence of VNTR alleles defined as fixed bins. Genetics *130*:873-887.

107. Weir, B. 1992. Population genetics in the forensic DNA debate. Proc. Natl. Acad. Sci. USA *89*:11654-11659.

108. Weir, B. 1994. The effects of inbreeding on forensic calculations. Ann. Rev. Genet. *28*:597-621.

109. Woller, J., B. Budowle, S. Furedi, and Z. Padar. 1996. Hungarian population data on the loci HLA-DQα, LDLR, GYPA, HBGG, D7S8, and Gc. Int. J. Legal Med. *108*:280-282.

110. Woller, J., B. Budowle, M. Angyal, S. Furedi, and Z. Padar. 1998. Population data on the loci HLA-DQA1, LDLR, GYPA, HBGG, D7S8, GC, and D1S80 in a Hungarian Romany population, p. 381-383. *In* B. Olaisen, B. Brinkmann, and P.J. Lincoln. (Eds.), Advances in Forensic Haemogenetics 7. Elsevier, Amsterdam.

111. Woo, K.M. and B. Budowle. 1995. Korean population data on the PCR-based loci LDLR, GYPA, HBGG, D7S8, Gc, HLA-DQA1, and D1S80. J. Forensic Sci. *40*:645-648.

Appendix III: Reagents and Supplies

General instructions are applicable in the preparation of all reagents:

1) Use graduated cylinders or pipets closest in capacity to the volume being measured for preparing liquid reagents.
2) Store all reagents in sterile containers unless otherwise noted. Label all reagents with name of reagent, date prepared, and initials of pre-parer.
3) Record each preparation in a Reagent Log Book.

REAGENTS AND SUPPLIES

ACES™ Human DNA Quantification Probe Plus Kit   Life Technologies (GIBCO BRL), Rockville, MD, USA; Catalog No. 0352-011

ACETIC ACID STOP SOLUTION   **10% Acetic Acid, 10 L** (CH_3COOH); Baker Chemicals, Phillipsburg, NJ, USA; Baker Analyzed grade, Catalog No. 9508-05, 99.8%

Combine 1 L glacial acetic acid with 9 L deionized H_2O. Mix thoroughly. Store at room temperature.

ACRYLAMIDE STOCK SOLUTION   **30% Total Acrylamide/2% Cross Linker Acrylamide;** Life Technologies; Catalog

No. 5512UC or Bio-Rad Laboratories,
Hercules, CA, USA; Catalog No.
161-0123
piperazine diacrylamide; Bio-Rad
Laboratories; Catalog No. 161-0202

Weigh out 29.4 g acrylamide and 0.6 g piperazine diacrylamide and add
both solids to approximately 70 mL deionized water while stirring. When the
solids have dissolved, bring the final volume to 100 mL with deionized
water. Store the solution at 4°C. Acrylamide and piperazine diacrylamide
should be electrophoretic grade.

> **CAUTION: Acrylamide and bis-acrylamide are neurotoxins. Avoid
> inhalation and skin contact. Wear a mask and gloves when weighing out
> acrylamide. Gloves should be worn when handling acrylamide solutions.**

**ACRYLAMIDE STOCK
SOLUTION**

**40% Acrylamide/5% Cross Linker,
19:1 Acrylamide:Bis-acrylamide**

Dissolve a 30 g bottle of acrylamide/bis (Bio-Rad Laboratories; Catalog No.
161-0123) in 48 mL ddH$_2$O while stirring. Add 1 g of Sigma amberlite MB-
1A and continue to stir for 30 min. Filter through a 0.2 μm filter. Adjust
final volume to 100 mL with ddH$_2$O. Store at 4°C for up to two weeks.

> **CAUTION: Acrylamide and bis-acrylamide are neurotoxins. Avoid
> inhalation and skin contact. Wear a mask and gloves when weighing out
> acrylamide. Gloves should be worn when handling acrylamide solutions.**

AG® 501-X8 (D) Resin

Bio-Rad Molecular Biology Grade;
Bio-Rad Laboratories; Catalog No.
143-6425

ALCONOX®

VWR Scientific Products, West Chester,
PA, USA; Catalog No. 21835-032 1104

**AMBERLITE MB-1
(Mixed Bed Exchanger)**

Sigma Chemical, St. Louis, MO, USA;
Catalog No. MB-1A.

Store at room temperature.

AMMONIUM ACETATE

7 M Ammonium acetate; Sigma
Chemical; Catalog No. A 7330

Dissolve 53.96 g anhydrous NH$_4$OAc and bring to 100 mL with water. Sterilize by suction through a sterile 0.45 μ filter.

**AMMONIUM PERSULFATE
(NH$_4$Cl)$_2$S$_2$O$_8$**

10% Ammonium persulfate; Sigma
Chemical; Catalog No. A- 3678 or
Bio-Rad Laboratories, Catalog No.
161-0700

Weigh out 0.1 g ammonium persulfate and dissolve in 1 mL sterile ddH$_2$O.

AMPFlSTR® PROFILER PLUS™ PCR AMPLIFICATION KIT

PE Biosystems, Foster City, CA, USA;
Catalog No. 4303326

AmpFlSTR Profiler Plus PCR Amplification Kit contains reagents necessary to co-amplify 100 reactions of the repeat regions of nine STR loci (D3S1358, vWA, FGA, D8S1179, D21S11, D18S51, D5S818, D13S317, D7S820) and the amelogenin locus. The kit includes: AmpFlSTR PCR Reaction Mix, AmpFlSTR Profiler Plus Primer Set, AmpliTaq™ Gold DNA Polymerase, AmpFlSTR Control DNA 9947A, Mineral oil, and Profiler Plus Allelic Ladder. Store each component as directed by the manufacturer. See Allelic Ladder (below) for preparation and storage conditions.

AMPFlSTR® COFILER™ PCR AMPLIFICATION KIT

PE Biosystems; Catalog No. 4305246

AmpFlSTR Cofiler PCR Amplification Kit contains reagents necessary to co-amplify 100 reactions of the repeat regions of six STR loci (D3S1358, D16S539, THO1, TPOX, CSF1PO, D7S820) and the amelogenin locus. The kit includes: AmpFlSTR PCR Reaction Mix, AmpFlSTR Cofiler Primer Set, AmpliTaq Gold DNA Polymerase, AmpFlSTR Control DNA 9947A, Mineral oil, and Cofiler Allelic Ladder. Store each component as directed by the manufacturer. See Allelic Ladder for preparation and storage conditions.

AMPFlSTR PCR Reaction Mix   1.1 mL/tube

The reaction mix contains two tubes each containing $MgCl_2$, dATP, dCTP, dGTP, dTTP, BSA, and 0.05% sodium azide in buffer and salt. Store at 4°C.

AMPFlSTR PROFILER PLUS PRIMER SET   1.1 mL

The primer set consists of one tube of locus-specific 5-FAM-, JOE-, and NED-labeled and unlabeled primers in buffer. Store at 4°C, protected from light.

AMPFlSTR COFILER PRIMER SET   1.1 mL

The primer set consists of one tube of locus-specific 5-FAM-, JOE-, and NED-labeled and unlabeled primers in buffer. Store at 4°C, protected from light.

AMPLITAQ GOLD DNA POLYMERASE   50 µL/tube

AmpliTaq Gold is supplied at an enzyme concentration of 5 U/µL. Store at -20°C.

AMPFlSTR CONTROL DNA 9947A   0.3 mL

The female human cell line DNA is supplied at a concentration of 0.1 ng/µL in 0.05% sodium azide and buffer. Store at 4°C.

AmpFlSTR PROFILER PLUS   Allelic Ladder, 50 μL
ALLELIC LADDER

A single tube contains the AmpFlSTR Blue™ (FGA, vWA, D3S1358), AmpFlSTR Green™ II (Amelogenin, D18S51, D21S11, 8S1179), and AmpFlSTR Yellow™ (D7S820, D13S317, D5S818) Allelic Ladders. Store at 4°C, protected from light.

AMPFlSTR COFILER   Allelic Ladder, 50 μL
ALLELIC LADDER

A single tube contains the Blue (D16S539, D3S1358), AmpFlSTR Green™ I (Amelogenin, CSF1PO, TPOX, THO1), and Yellow (D7S820) Allelic Ladders. Store at 4°C, protected from light.

AMPFlSTR® COFILER AND   PE Biosystems; Catalog No. 4305979
PROFILER PLUS™ PCR
AMPLIFICATION KIT

This is a combination kit containing the reagents from both the Cofiler and Profiler Plus kits.

AMPLITYPE HLA DQ ALPHA   PE Biosystems; Catalog No. N801-0056
FORENSIC DNA TYPING KIT

The kit contains reagents necessary to amplify and type the DQα gene. These reagents, enough for 50 reactions, include: HLA DQα PCR Reaction Mix, 8 mM $MgCl_2$ solution, mineral oil, control DNA, HLA DQα Probe Strips, PCR Reaction Tubes, enzyme conjugate (horseradish peroxidase-streptavidin conjugate), and chromogen (3,3′,5,5′-tetramethylbenzidine). Store each component as directed by the manufacturer. See chromogen (below) for preparation and storage instructions.

AMPLITYPE PM PCR   PE Biosystems; Catalog No. N808-0057
AMPLIFICATION AND
TYPING KIT

The kit contains reagents necessary to amplify and type six loci (DQα, LDLR, GYPA, HBGG, D7S8, and GC). The kit includes: AmpliType PM PCR Reaction Mix, primers, control DNA, mineral oil, AmpliType PM DNA Probe Strips, enzyme conjugate (horseradish peroxidase-streptavidin conjugate), and chromogen (3,3′,5,5′-tetramethylbenzidine). Store components as directed by the manufacturer. See chromogen for preparation and storage instructions.

AMPLITYPE DQα   50 μL Aliquots
REACTION MIX

Upon receipt of the AmpliType DQα kit, remove the PCR Reaction Mix bottle. Carefully aliquot 50 μL PCR mix into each of the reaction tubes provided using a repeater pipet dedicated solely to this purpose. This must be done in a room free of amplified DNA. Place PCR tubes in a rack or box not used for DNA preparation or amplified DNA handling. Store tubes, separated from any source of DNA, at 4°C. Do not freeze. Discard after expiration date stated with kit.

AMPLITYPE PM REACTION MIX

40 μL Aliquots

The Reaction Mix contains 6 mM $MgCl_2$, AmpliTaq DNA Polymerase (*Taq* polymerase), dATP, dCTP, dGTP, and dTTP in buffer, and salt. When ready to use the AmpliType PM Kit, remove the bottle of Reaction Mix and aliquot 40 μL into autoclaved tubes.

> **CAUTION: This step must be performed either in a biological hood or in a room free from amplified DNA.**

Cap tubes tightly. Place tubes in a rack not used for DNA preparation or amplified DNA handling. Store tubes separated from any source of DNA at 4°C. Discard after expiration date stated with kit.

AMPLITYPE PM + DQA1 PCR AMPLIFICATION AND TYPING KIT

PE Biosystems; Catalog No. N808-0094

The kit contains reagents necessary to amplify and type six loci (DQA1, LDLR, GYPA, HBGG, D7S8, and GC). The kit includes: AmpliType PM PCR Reaction Mix, primers, control DNA, mineral oil, AmpliType DQA1 and PM DNA Probe Strips, enzyme conjugate (horseradish peroxidase-streptavidin conjugate), and chromogen (3,3',5,5'-tetramethylbenzidine). Store components as directed by the manufacturer.

AMPLITAQ DNA POLYMERASE

PE Biosystems; Catalog No. N801-0060

The enzyme is supplied at an enzyme concentration of 5 U/μL. Store at -20°C.

GENEAMP™ 10× PCR Buffer

PE Biosystems, Catalog No. N808-006

Buffer contains: 100 mM Tris-HCl, pH 8.3, 500 mM KCl, 15 mM $MgCl_2$, and 0.01% (wt/vol) gelatin. Store at -20°C.

ANODE BUFFER JAR

Vial for 1× buffer; PE Biosystems; Catalog No. 005402

10× GENETIC ANALYZER BUFFER WITH EDTA

PE Biosystems; Catalog No. 402824

1× GENETIC ANALYZER BUFFER

1× Buffer, 15 mL

Combine 1.5 mL 10× Genetic Analyzer buffer with 13.5 mL filter-purified water. This volume is sufficient to set up one CE instrument and can be used for 48 hours or 100 injections (whichever comes first). Or it can be stored at 4°C for 2 weeks.

BASKET INSERT	**SPIN-X insert, no membrane;** Costar; Fisher Scientific, Pittsburgh, PA, USA; Catalog No. 9301
BESSMAN TISSUE PULVERIZER	Fisher Scientific; Catalog No. 08-418-3
BLEACH SOLUTION	**10% sodium hypochlorite solution, 100 mL**

Bring 10 mL household bleach to 100 mL with deionized water. Store at room temperature.

BLUE DEXTRAN/EDTA	PE Biosystems; Catalog No. 402055
BORIC ACID (H_3BO_3)	Sigma Chemical; Catalog No. B-6768, molecular biology grade
BOVINE SERUM ALBUMIN (BSA)	**1.6 µg/µL Bovine Serum Albumin** Sigma Chemical; Catalog No. A3550

Add 16 mg bovine serum albumin to 10 mL of sterile ddH_2O. Gently stir until the BSA is dissolved. Adjust final volume to 10 mL with sterile ddH_2O to achieve a concentration of 1.6 mg/mL. Aliquot 100-µL portions and store at -20°C.

BUFFER VIALS	**4 mL Buffer Vials with Cap Adapters** PE Biosystems; Catalog No. 401955
CAPILLARY 5/pkg (STR typing)	**310 Capillaries, 47 cm × 50 µm;** PE Biosystems; Catalog No. 402839
CAPILLARY (mtDNA quantification)	Alltech Associates, Deerfield, IL, USA; Catalog No. 93567 J&W DB 17 100 cm, 50 µm
CENTRI-SEP SPIN COLUMNS	Princeton Separations, Adelphia, NJ, USA; Catalog No. 00105
CHELEX SOLUTION	**approx. 5% wt/vol, 100 mL;** 100-200 mesh, sodium form, Bio-Rad Laboratories; Catalog No. 143-2832

Weigh out 5 g Chelex 100 Resin into a bottle that has been autoclaved with a stir bar in it. Add 100 mL sterile water. Store at room temperature.

CHELEX SOLUTION	**approx. 20% wt/vol, 100 mL;** 100-200 mesh, sodium form, Bio-Rad Laboratories; Catalog No. 143-2832

Weigh out 20 g Chelex 100 Resin into a bottle that has been autoclaved with a stir bar in it. Add 100 mL sterile water. Store at room temperature.

Note: Gently mix Chelex solutions with stir bar when pipetting.

CHROMOGEN SOLUTION   30 mL

Bring the bottle containing 60 mg chromogen [3,3',5,5'-tetramethylbenzi-dene, (TMB)] provided with the AmpliType kit to room temperature. Before opening the bottle, tap bottle on the lab bench to shake the chromogen to the bottom of the bottle. Carefully remove the stopper to prevent loss of powder. Reconstitute by adding 30 mL 100% ethanol (reagent grade, room temperature).

Note: Do not use ethanol that has been stored in a metal container. Do not use 95% ethanol. Recap the bottle. Seal the bottle with Parafilm M™. Agitate bottle to remove any chromogen from the stopper. Store at 2–8°C in the brown bottle in which the powdered TMB came. This solution is stable for four months after reconstitution.

CITRATE BUFFER

0.1 M Sodium Citrate, pH 5.0, 5 L
$Na_3C_6H_5O_7 \cdot 2H_2O$; Sigma Chemical; Catalog No. C-8532 molecular biology grade; $C_6H_8O_7$, free acid, anhydrous, Sigma Chemical; Catalog No. C-0759

Dissolve 92.0 g trisodium citrate dihydrate in 4 L deionized water. Adjust pH to 5.0 by addition of citric acid. It will take approximately 37.5 g of the free acid to bring the pH to 5.0. Continuously monitor the pH as the free acid is being added. Adjust final volume to 5 L using deionized water. Store at room temperature.

COMB (24-well sharks-tooth)

ABD PE Biosystems; Catalog No. 401048

CONVERTIBLE® FILTRATION and MANIFOLD SYSTEM
(Slot blot Apparatus)

Life Technologies; Catalog No. 11055-068, **48-well Slot Top Plate Gasket**, Baseplate, Catalog No. 11055-019

CORNING CELLULOSE ACETATE FILTER

.45 μm low protein binding membrane
VWR; Catalog No. 25 943-500

COTTON SHEETING   Washed, cotton sheeting

100% cotton bedsheets (180 threads/inch) are washed once with detergent in hot water, followed by two washes in cold water without detergent. After air-drying, sections are removed and placed in boiling water for 20 minutes followed by air-drying. Store in zipper-type plastic bags at room temperature.

D1S80 ALLELIC REFERENCE LADDER

AmpliFLP D1S80 Allelic Ladder
PE Biosystems; Catalog No. N808-0064

D1S80 allelic reference ladder consists of 27 amplified D1S80 alleles. Store at 4°C.

DELRIN COMB	Life Technologies; Catalog No. 11092-095, 0.4 mm, 18 tooth

DNA PROBE FOR HUMAN QUANTITATION   **Human DNA Probe D17Z1**

D17Z1 DNA probe is supplied in the ACES™ Human DNA Quantification Probe Plus Kit (Life Technologies; Catalog No. 10352-011). Store at 4°C.

DNA STANDARD - K562 Cell Line   **DNA from K562 human cell line** Life Technologies; Catalog No. 14410-013

DNA Typing Grade™ K562 is used to prepare the size standards. This DNA is supplied at a concentration of 200 μg/mL. Dilute the supplied DNA with Tris-EDTA (TE), pH 8.0 to achieve a 20 ng/μL stock concentration. This stock is diluted to prepare the eight size standards (20 ng/μL, 10 ng/μL, 4 ng/μL, 2 ng/μL, 1 g/μL, 0.4 ng/μL, 0.2 ng/μL, 0.1 ng/μL). Store the supplied DNA at -20°C. Store the diluted DNA at 4°C.

DEOXYRIBONUCLEOSIDE-TRIPHOSPHATES   **GeneAmp dNTPs;** PE Biosystems; Catalog No. N808-0007

GeneAmp dNTPs contain dATP, dCTP, dGTP, and dTTP. Each deoxyribonucleoside triphosphate is present at a concentration of 10 mM. Store at -20°C.

DEVELOPMENT FOLDER   **Photogene™ Development Folder;** Life Technologies; Catalog No. 18195-016

Photogene™ development folders are used to encase the slot blot membranes before they are loaded into cassettes. Store at room temperature.

DNA SEQUENCING KIT   **dRhodamine Terminator Cycle Sequencing Ready Reaction Kit;** ABI PRISM™, PE Biosystems; Catalog No. 403042

Prior to use, allow the kit to thaw to room temperature and briefly centrifuge. Store at -20°C.

DEOXYRIBONUCLEOSIDE-TRIPHOSPHATES [(dNTPs) GeneAmp™ dNTPs]   **GeneAmp dNTPS;** PE Biosystems; Catalog No. N808-0007

Contains dATP, dCTP, dGTP, and dTTP. Each dNTP is present at a concentration of 10 mM. Store at -20°C.

DNA QUANTIFICATION STANDARDS   **Lambda phage DNA at 250 μg/mL = stock solution**

For use on test gels

Carry out serial doubling dilutions of the stock with TE buffer to obtain the solutions shown. Combine 1.0-mL aliquots of diluted standards with loading solution as shown.

1.0 mL at 125.0 μg/mL + 0.5 mL loading solution = 500 ng/6 μL
1.0 mL at 62.5 μg/mL + 0.5 mL loading solution = 250 ng/6 μL
1.0 mL at 31.3 μg/mL + 0.5 mL loading solution = 125 ng/6 μL
1.0 mL at 15.6 μg/mL + 0.5 mL loading solution = 63 ng/6 μL
1.0 mL at 7.8 μg/mL + 0.5 mL loading solution = 31 ng/6 μL
1.0 mL at 3.9 μg/mL + 0.5 mL loading solution = 15 ng/6 μL

Note: Quantitative standards of DNA can be obtained commercially that provide a similar range of concentrations.

DREMEL® Motor Tool®	Model 395, Type 4, Dremel®; Racine, WI, USA

Dithiothreitol (DTT); $C_4H_{10}O_2S_2$	**1 M DTT, 10 mL;** Sigma Chemical; Catalog No. D-9779, molecular biology grade

Dissolve 1.54 g DTT in 10 mL sterile filter-purified water in a sterile, disposable plastic 15-mL tube. Do not autoclave. Aliquot portions of solution and store at -20°C. Discard any unused portion of a thawed tube.

EDTA	**50 mM, pH 8.0** $C_{10}H_{14}N_2O_8Na_2 \cdot 2H_2O$, Sigma Chemical; Catalog No. E-5134, molecular biology grade

For a 100 mL stock solution, slowly add 1.86 g ethylenediaminetetraacetic disodium salt, dihydrate to 90 mL ddH$_2$O. Stir vigorously on a magnetic stirrer. Adjust to pH 8.0 by adding 10 N NaOH.

Note: The EDTA at this concentration will not readily go into solution until the pH is adjusted to 8.0. Adjust final volume to 100 mL with ddH$_2$O. Autoclave. Store at room temperature.

EDTA	**0.5 M, pH 8.0**

For a 1 L stock solution, slowly add 186.1 g ethylenediaminetetraacetic disodium salt, dihydrate to 800 mL ddH$_2$O. Stir vigorously on a magnetic stirrer. Adjust to pH 8.0 by adding 10 N NaOH. Note: The EDTA at this concentration will not go into solution until the pH is adjusted to 8.0. Adjust final volume to 1 L with ddH$_2$O. Autoclave. Store at room temperature.

EMERY DISKS	**Economy cutting disks**, #2 Medium, Henry Schein, Melville, NY, USA; Catalog No. 100-6804
ENDODONTIC FILES	**21-mm K-files**, 15-80, Henry Schein; Catalog No. 100-9709, 100-1215, 100-4863, and 100-5087
ETHANOL (EtOH)	**Absolute**

| ETHIDIUM BROMIDE | 5 mg/mL; Sigma Chemical; Catalog No. E-7637, molecular biology grade |

Dissolve 250 mg 2,7-diamino-10-ethyl-9-phenylphenanthridinium in approximately 45 mL deionized water. Adjust volume to 50 mL. Store at 2–8°C protected from light (e.g. in a brown bottle covered with aluminum foil).

> **Warning: Ethidium Bromide is a mutagen. Always wear gloves when handling it.**

| FILM
KODAK X-OMAT™ AR | Kodak Scientific Imaging
XAR-5 Film; Kodak; Catalog No. 165 1454, 20.3 × 25.4 cm, 8 × 10″, 50/91.15 |

| FINAL WASH BUFFER
10× ACES 2.0 Final Wash | Life Technologies; Catalog No. 10355-014 |

BRL ACES 2.0 10× Final Wash Buffer is supplied in a 10× concentration and must be diluted to a 1× concentration for use. Store at 4°C.

| 1× FINAL WASH BUFFER | BRL ACES 2.0 |

Combine 50 mL of 10× Final Wash Buffer with 450 mL of filter-purified water. Store at room temperature.

| ELECTRODE (CAPILLARY) | Platinum Cathode Electrode; PE Biosystems; Catalog No. 005914 |

| FLUSH-CUTTING WIRE CUTTERS | Electrode trimmers; PE Biosystems; Catalog No. T-6157 |

| FORMAMIDE | Formamide, Redistilled, 500 g
Amresco, Solon, OH, USA;
Catalog No. 0606 |

1.0-mL aliquots are stored at -20°C in screw-capped tubes. Aliquots can be stored at 4°C for two weeks.

> **WARNING: Formamide is a teratogen. Avoid inhalation, skin contact, or ingestion. Use gloves when handling. Dispose of unused portions properly in appropriate hazardous waste containers.**

| FORMAMIDE/GS-500 [ROX] | Formamide/GS-500 [ROX] Master Mix |

The formamide/GS-500 [ROX] solution can be prepared as a master mix by combining formamide with GS-500 [ROX] sizing standard in a ratio of 74 formamide to 1 GS500. Store at 4°C for two weeks.

> **WARNING: Formamide is a teratogen. Avoid inhalation, skin contact or ingestion. Use gloves when handling. Dispose of unused portions properly in appropriate hazardous waste containers.**

GELBOND PAG FILM   FMC Bioproducts; Catalog No. 54735,
                           195 mm • 370 mm

Wear clean gloves when handling GelBond.

GLASS PIPETTES   Cotton-plugged Pasteur pipettes, 9″
                           Fisher Scientific; Catalog No.
                           13-678-6B

GLASS PLATES   Gel preparation plates
                           Life Technologies; Catalog No. 1093KG
                           (sold as a pair)

Plates come in the following dimensions:

  Short plate dimensions: 19.6 cm × 33.8 cm × 0.5 cm

  Long plate dimensions: 19.6 cm × 36.8 cm × 0.5 cm

GLYCEROL WASH   5% (vol\vol) Glycerol, 5 L
                           $CH_2OHCHOHCH_2OH$; ACS grade,
                           99.5% Fisher Scientific; Catalog No.
                           G33-4

Add 250 mL glycerin to 4 L deionized H_2O. Mix thoroughly and bring to a
final volume of 5 L with deionized H_2O. Store at room temperature.

HEPES-BUFFERED SALINE N-2-hydroxyethylpiperazine-N′-2-
                           ethanesulfonic acid; Sigma Chemical;
                           Catalog No. H1016

Dissolve 2.383 g HEPES and 8.415 g NaCl in 900 mL water. Titrate to pH
7.2 with NaOH. Bring to a final volume of 1.0 L with water.

HERRING SPERM DNA   500 mg herring sperm DNA
                           Sigma Chemical; Catalog No. D 3159

Bring to a final volume of 50 mL with water.

HYBRIDIZATION SOLUTION BRL ACES 2.0 Hybridization Buffer
                           Life Technologies; Catalog No.
                           14271-019

BRL ACES 2.0 Hybridization Buffer is warmed to 50°C before hybridization. Store at 4°C.

HYBRIDIZATION SOLUTION 5× SSPE - 0.5% wt/vol SDS, 1 L
                           (DQA1/PM)

Mix together 250 mL 20× SSPE, 25 mL 20% wt/vol SDS, and 725 mL deionized water. Store at room temperature. Solids must be in solution before use;
warming may be necessary.

HYDROCHLORIC ACID   Fisher Scientific; Catalog No. A-144,
(concentrated)   Reagent ACS grade, 36.5–38.0%

1% HEC IN BUFFER	**100 mM Tris-100 mM Boric acid-2 mM EDTA**
	Trizma base; Sigma Chemical; Catalog No. T-8524
	Boric acid; Sigma Chemical; Catalog No. B-6768
	HEC; Aldrich; Catalog No. 30,863-3
	Corning cellulose acetate low protein binding filter; VWR; Catalog No. 161-00451

To make a 500-mL stock solution, dissolve 6.055 g Tris base, 3.0915 g Boric acid, and 0.2925 g EDTA in 400 mL ddH$_2$O. Slowly add 5 g hydroxyethyl cellulose while stirring. Vigorously mix the solution. (HEC requires approximately 2 hours to go into solution). Bring the volume to 500 mL with ddH$_2$O, and vacuum filter buffer solution through a .45 μm corning cellulose acetate low protein binding filter. pH should be between 8.1–8.5. Can be stored at room temperature.

1% HEC rinse/separation buffer is prepared by adding YO-PRO dye to a final concentration of 50 ng/mL. It should be noted that YO-PRO dye is light sensitive and should be stored in a manner to eliminate exposure to light.

INTERNAL LANE STANDARD	**Prism GeneScan-500 (ROX), 400 μL**
	PE Biosystems; Catalog No. 401734

Store at 4°C.

LOADING SOLUTION	**5:1 Deionized formamide/blue**
Sequencing	**dextran/EDTA**
	Formamide; IBI; Catalog No. 1B72020

For each sample to be loaded, mix 5 parts deionized formamide and 1 part blue dextran/EDTA in a sterile 1.5-mL tube.

CAUTION: Formamide is an embryotoxin. This substance should be used in a chemical hood. Precautions should be taken to avoid skin contact.

LOADING SOLUTION	**50% glycerol - 0.1% bromphenol blue -**
for submarine agarose gels	**0.1 M EDTA in TE, 100 mL**
	bromphenol blue; Sigma Chemical; Catalog No. B-5525

Pour 50 mL glycerol into a 100-mL graduated cylinder. Add 20 mL 0.5 M EDTA, 30 mL TE, and 0.1 g 3′,3″,5′,5″-tetrabromophenolsulfonephthalein (bromphenol blue).

LOADING SOLUTION	**50% Sucrose - 0.25% Bromophenol**
for native polyacrylamide gels	**blue in 240 mM Formate-Tris buffer**

Dissolve 25 μg bromophenol blue and 5 g sucrose in a final volume of 10 mL of 240 mM Tris-Formate buffer. Store at 4°C.

LOADING SOLUTION
for sequencing gels

5:1 Deionized formamide/50 mM EDTA
formamide; IBI; Catalog No. 1B72020

For each sample to be loaded, mix 5 parts deionized formamide and 1 part 50 mM EDTA, pH 8.0 in a sterile 1.5-mL tube.

> **CAUTION: Formamide is an embryotoxin. This substance should be used in a chemical hood. Precautions should be taken to avoid skin contact.**

LUMI-PHOS® PLUS

Lumi-phos Plus Detection Reagent; Life Technologies; Catalog No. Y01946

Lumi-Phos Plus is photo sensitive and should be stored at 4°C.

Note: Life Technologies Catalog No. 10352-011 is the ACES Human DNA Quantification Probe Plus Kit and also contains DNA probe D17Z1.

MATRIX STANDARDS

Dye Primer Matrix Standards Kit
PE Biosystems; Catalog No. 401114

NED Matrix Standard; PE Biosystems; Catalog No. 402996

The Dye Primer Matrix Standards Kit and NED Matrix Standard contain reagents necessary to create a matrix file for use with Profiler Plus and Cofiler amplified samples. The standards required are 5-FAM (blue), JOE (green), and ROX (red) contained in the Dye Primer Matrix Standards Kit, and NED (yellow). NOTE: Do not use the TAMRA matrix standard provided in the kit.

METHANOL (CH_3OH)

HPLC GRADE; Fisher Scientific; Catalog No. A452-4

Store at room temperature.

MICROAMP™ RETAINER

PE Biosystems; Catalog No. N801-0530

MICROCENTRIFUGE TUBES

2.2 mL, Costar; Catalog No. 3214

MICROCENTRIFUGE TUBES

0.5 mL, Sarstedt; Catalog No. 72.699

MICRO TISSUE GRINDER

Kontes, Tissue Grinder, Micro 88540-0000; Fisher Scientific; Catalog No. K885470-0000

MICROCON™ 100

Amicon, Beverly, MA, USA; Catalog No. 42413

NALGENE® FILTER

Tissue Culture Filter Unit .2 μm cup filter; VWR; Catalog No. 28199-621

| Nitric Acid | 1% HNO$_3$, 10 L; Sigma Chemical Catalog No. 25,811-3 |

Add 100 mL concentrated nitric acid (69%–71%) to 9 L deionized H$_2$O. Mix and bring to a final volume of 10 L with deionized H$_2$O. Store at room temperature.

CAUTION: Nitric acid is a strong oxidizer, is corrosive, and will cause severe burns. Wear gloves and safety glasses when preparing this solution.

| NUSEIVE 3:1 AGAROSE | FMC, Rockland, ME, USA; Catalog No. 50090 |

| PANAVISE® | PanaVise Products Inc., Reno, NV, USA, Model 381 Vacuum base PanaVise |

| PCIA (Phenol/Chloroform/Isoamyl alcohol) | 25:24:1 (vol\vol); Life Technologies; Catalog No. 5593UB or Sigma Chemical; Catalog No. P3803 |

PCIA saturated with buffer to a pH of 7.9 is used for the purification of DNA.

CAUTION: This solution is an irritant and is toxic. Its use should be confined to a fume hood. Gloves and a mask should be worn.

PHOSPHATE BUFFERED SALINE (PBS)	2.7 mM KCl - 137 mM NaCl- 1.5 mM KH$_2$PO$_4$ - 8.0 mM Na$_2$HPO$_4$, pH 7.4, 5 L KCl; Sigma Chemical; Catalog No. P-3911, ACS reagent or S- 9541, molecular biology grade
	sodium chloride NaCl; Sigma Chemical; Catalog No. S-3014, molecular biology grade
	potassium phosphate monobasic KH$_2$PO$_4$, anhydrous; Sigma Chemical; Catalog No. P-0662, ACS reagent
	anhydrous disodium phosphate sodium phosphate dibasic, Na$_2$HPO$_4$; Sigma Chemical; Catalog No. S- 0876 or S-3264, molecular biology grade

Dissolve 1 g potassium chloride (KCl), 40 g sodium chloride (NaCl), 1 g potassium phosphate monobasic (KH$_2$PO$_4$, anhydrous), and 5.5 g anhydrous disodium phosphate (sodium phosphate dibasic, Na$_2$HPO$_4$) in 4000 mL deionized water. Check pH, adjust to 7.4 if necessary with concentrated hydrochloric acid (HCl). Adjust final volume to 5 liters. Autoclave. Store at room temperature. Aliquot amount needed for appropriate number of extractions into a sterile, disposable plastic tube and discard unused portion of aliquot.

50% POLYETHYLENE GLYCOL (PEG)

Polyethylene Glycol (PEG, 8000 MW), 100 mL; Sigma Chemical; Catalog No. P-5413, molecular biology grade

Dissolve 50 g polyethylene glycol in approximately 50 mL deionized water. Adjust final volume to 100 mL. Store at room temperature.

POLYMER (CE)
for STR typing

Performance Optimized Polymer 4 (POP-4®), 5 mL; PE Biosystems; Catalog No. 402838

PRIMERS (Oligonucleotide DNA Primers)

Primers can be custom synthesized by a number of laboratories, for example Operon Technologies Inc., Alameda, CA or Life Technologies, Gaithersburg, MD. Synthesized primers are quantified by UV absorption at 260 nm and the quantity stated in picomoles. Lyophilized primers should be stored at -20°C. Each primer is reconstituted from the lyophilized form using ddH$_2$O to achieve a final concentration of 100 µM (100 nanomoles/mL). Storage stocks should consist of small aliquots stored at -20°C.

PRIMERS (D1S80)   **Single-Stranded DNA Primers**

Two single-stranded DNA primers are required for the amplification of alleles at locus D1S80. The sequences for the primers (shown from their 5′ end) are as follows:

D1S80A - 28 base oligonucleotide

5′- GAA ACT GCC TTC CAA ACA CTG CCC GCC G,

D1S80B - 29 base oligonucleotide

5′- GTC TTG TTG GAG ATG CAC GTG CCC CTT GC.

Primers as used in this procedure have been fully deprotected, desalted, and lyophilized. They have been custom synthesized by Operon Technologies Inc., Alameda, CA, and should be stored dry at -20°C until reconstitution. Synthesized primers are quantified by UV absorption at 260 nm and the quantity of lyophilized primer will be stated in pmoles. Each primer is reconstituted from the lyophilized form with Tris/EDTA, pH 7.5 to achieve a final concentration of 100 µM (100 nmoles/mL) as a storage stock solution. Reconstituted primers should be stored at -20°C.

PRIMER STORAGE STOCK SOLUTION

Volume of reconstitution = (pmoles primer supplied)/(100,000 pmoles/mL)

The stock primer solution must be diluted 1:8 to obtain the primer solution that is added to the amplification mixture. Combine 60 µL primer stock with 420 µL sterile, deionized H$_2$O to obtain the working primer solution. This must be done for each of the two primers. Store primers at -20°C.

PROTEINASE K (20 mg/mL)   AMRESCO; Catalog No. E195-25

Using a pipettor with a sterile tip, aliquot the liquid Proteinase K into 0.5-mL microcentrifuge tubes and store at 4°C. Discard any unused portions of opened tubes.

> **CAUTION: Powdered Proteinase K and solutions of Proteinase K can be irritating to mucous membranes. Wear safety glasses and gloves when handling.**

PUMP BLOCK   Genetic Analyzer Pump Block; PE Biosystems; Catalog No. 604072

QUANTIFICATION 200 BP STANDARD (100 ng/µL STANDARD) for CE/mtDNA quantitation   GenSura

REACTION TUBES (MicroAmp™ Reaction Tubes)   PE Biosystems; Catalog No. N801-0533

REACTION CAPS (MicroAmp Caps)   PE Biosystems; Catalog No. N801-0535

RETAINER CLIPS   Genetic Analyzer Retainer Clips (96-Tube Tray Septa Clips), 4/pkg PE Biosystems; Catalog No. 402866

20% SARKOSYL   **N-Laurylsarcosine, 100 mL** $C_{15}H_{28}NO_3Na$, Sigma Chemical; Catalog No. L5125

Add 20 g N-laurylsarcosine to filter-purified H_2O and stir until dissolved. Bring to a final volume of 100 mL with filter-purified H_2O and sterilize by passage through a 0.45 µ sterile filter.

SEPTA   Genetic Analyzer Septa Strips for 96-Tube Tray, 485/pkg (20 strips) PE Biosystems; Catalog No. 402059

                                Genetic Analyzer Septa Strips for 0.5-mL Sample Tubes; PE Biosystems; Catalog No. 401956

Silver Nitrate   **0.2% AgNO₃;** Sigma Chemical; Catalog No. S0139

Dissolve 1 g AgNO₃ in 500 mL deionized H_2O. Prepare at least 15 minutes before time of intended use. The solution should be used during the day of preparation. Silver nitrate is highly toxic and light sensitive. Wear gloves when handling.

SLOT BLOT PROBE   D17Z1, Oncor, Gaithersburg, MD, USA;
                                       Catalog No. P5040

The probe used for slot blot quantification of DNA, D17Z1

SLOT BLOT WASH BUFFER   **2× SSC - 0.5% SDS, 1 L**

Combine 100 mL 20× SSC, 25 mL 20% SDS, and 875 mL deionized water.
Store at 65°C.

SLOT BLOT HYBRIDIZATION   Life Technologies, Catalog No.
BUFFER   14271-019
(ACES 2.0 Hyb. Sol. DNA
Typing Grade)

SLOT BLOT MEMBRANE   **Pall Biodyne™ A Membrane;** Life
                               Technologies; Catalog No. M866-016

Pall Biodyne A membrane is amphoteric. Store at room temperature.

SLOT BLOT MEMBRANE   **0.2 M Tris-HCl - 2× SSC**
RINSE BUFFER

To make a 1 L stock solution, combine 200 mL 1.0 M Tris-HCl, pH 8.0, 100
mL 20× SSC, and 700 mL ddH$_2$O. Store at room temperature.

SLOT BLOT PROBE   Life Technologies; Catalog No.
(ACES Human DNA   4270SA
Quantitation Probe Plus)

Store at 4°C.

SODIUM CARBONATE -   **280 mM SODIUM CARBONATE**
FORMALDEHYDE -   **0.5% (vol\vol) FORMALDEHYDE, 10 L**
DEVELOPING SOLUTION   **anhydrous sodium carbonate;**
for silver staining   **Na$_2$CO$_3$,** Sigma Chemical; Catalog No.
                               S6139; **formaldehyde; CH$_2$O,** Sigma
                               Chemical; Catalog No. F1635

Add 296.8 g of anhydrous sodium carbonate to 9 L deionized H$_2$O. Stir until
dissolved and bring to a final volume of 10 L with deionized H$_2$O. Add 5
mL of 37% formaldehyde to the sodium carbonate solution and mix thoroughly. This solution is stable for one week. Store at room temperature.

**CAUTION: Sodium carbonate is an irritant and formaldehyde is highly
toxic.**

SODIUM CHLORIDE   **5.0 M NaCl;** Sigma Chemical; Catalog
                               No. S-3014

To make a 1 L stock solution, dissolve 292.2 g sodium chloride in approximately 700 mL ddH$_2$O. Adjust volume to 1 L with ddH$_2$O. Autoclave. Store
at room temperature.

SODIUM HYDROXIDE

10 N NaOH, 1 L
pre-made; Fisher Scientific; Catalog
No. SS255; **pellets;** Fisher Scientific;
Catalog No. S318

To make a 1 L stock solution, use a pre-made 10 N sodium hydroxide, or prepare as follows: dissolve 400 g sodium hydroxide pellets in approximately 700 mL ddH$_2$O. Adjust volume to 1 L with ddH$_2$O. Store at room temperature.

NaOH (4.0 M)

Dissolve 800 g NaOH pellets in approximately 4.2 L water. Bring to a final volume of 5.0 L with water.

SODIUM CHLORIDE/SODIUM HYDROXIDE

0.5M NaCl-0.5N NaOH

To make a 100 mL stock solution, combine 5 mL 10 N NaOH, 10 mL 5.0 M NaCl, and 85 mL ddH$_2$O. Or combine 100 mL 10 N NaOH, 200 mL 5.0 M NaCl, and 1700 mL filter-purified water. Autoclave. Store at room temperature.

20× SSC

3.0 M NaCl - 300 mM
C$_6$H$_5$O$_7$Na$_3$·2H$_2$O, pH 7.0, 1 L
NaCl; Sigma Chemical; Catalog No.
S-301, molecular biology grade
citric acid trisodium salt, dihydrate;
C$_6$H$_5$O$_7$Na$_3$·2H$_2$O; Sigma Chemical;
Catalog No. 8532, molecular biology
grade

Add 175.3 g sodium chloride and 88.2 g citric acid trisodium salt, dihydrate (C$_6$H$_5$O$_7$Na$_3$·2H$_2$O) to approximately 800 mL filter-purified water. Adjust the pH to 7.0 with concentrated hydrochloric acid (HCl). Adjust the final volume to 1 L using filter-purified water. Autoclave. Store at room temperature.

2× SSC

300 mM NaCl - 30 mM
C$_6$H$_5$O$_7$Na$_3$·2H$_2$O, pH 7.0, 1 L

Add 100 mL 20× SSC to 900 mL filter-purified water. Store at room temperature.

0.1× SSC : 0.5% SDS

Mix 5 mL 20× SSC and 25 mL 20% SDS. Bring to a final volume of 1.0 L with water.

2× SSC : 0.1% SDS

Mix 100 mL 20× SSC and 5 mL 20% SDS. Bring to a final volume of 1.0 L with water.

0.1× SSC : 0.1% SDS

Mix 5 mL 20× SSC and 5 mL 20% SDS. Bring to a final volume of 1.0 L with water.

20× SSPE BUFFER **3.6 M NaCl - 200 mM NaH₂PO₄ - 20 mM EDTA, pH 7.4, 2 L**
NaCl; Sigma Chemical; Catalog No. S-3014, molecular biology grade
sodium phosphate monobasic, NaH₂PO₄; Sigma Chemical; Catalog No. S-0751 or S-3139, molecular biology grade

Add 420.8 g sodium chloride and 48.0 g anhydrous sodium phosphate to approximately 1800 mL deionized water. When dissolved, add 80 mL 0.5 M Na₂EDTA·2H₂O and stir until mixed. Adjust the pH to 7.4 with 10 N sodium hydroxide (NaOH). Adjust the final volume to 2 liters using deionized water. Autoclave. Store at room temperature.

SODIUM DODECYL SULFATE **SDS, 1 L (CH₃(CH₂)₁₁OSO₃Na)**
SDS -20% (wt/vol) **Lauryl Sulfate, molecular biology grade;** Sigma Chemical; Catalog No. L4390 or **electrophoresis grade (ultra-pure);** Roche Molecular Products, Indianapolis, IN, USA; Catalog No. 000155

Slowly dissolve 200 g SDS in 800 mL filter purified water. Solution may be heated to facilitate the SDS to go into solution. Adjust volume to 1 L. Store at room temperature.

CAUTION: SDS is a respiratory irritant, wear a mask when working with powdered SDS.

SPERM WASH BUFFER **10 mM Tris-HCl - 10 mM EDTA - 50 mM NaCl - 2% SDS, pH 8.0, 500 mL**

Add 5 mL 1.0 M Tris-HCl, pH 8.0, 10 mL 0.5 M EDTA, pH 8.0, 5.0 mL 5 M NaCl, and 50 mL 20% SDS to 430 mL filter-purified water and adjust volume to 500 mL. Check pH. Autoclave. Store at room temperature.

SPERMINE Sigma Chemical; Catalog No. S 2876

Dissolve 348.2 mg Spermine-4HCl and bring to 10 mL with water.

Spin-EASE™ Extraction Tubes Life Technologies; Catalog No. 10238-012

SPOON EXCAVATORS **33L**, Schein Dental, Melville, NY, USA; Catalog No. 600-2398

SEB (STAIN EXTRACTION BUFFER)	**10 mM TRIS - 100 mM NaCl - 39 mM DTT - 10 mM EDTA - 2% SDS**

To make a 1 L stock solution, dissolve 5.84 g NaCl in 500 mL sterile ddH_2O. Add 10 mL 1.0 M Tris, 20 mL 0.5 M EDTA, and 100 mL 20% SDS. Titrate to pH 8.0 with HCl. Bring final volume to 1 L with ddH_2O. Store at room temperature.

Supplement with DTT before use. To 100 mL of the above solution, add 601.4 mg DTT and stir until dissolved. Store at room temperature. The complete solution is good for no more than two weeks.

STRYKER SAW	**Stryker 810 Autopsy Saw**, Stryker Inc., Kalamazoo, MI, USA
SULFURIC ACID	**1 N H_2SO_4;** Fisher Scientific; Catalog No. SA212-1
SULFURIC ACID, FUMING	Fisher Scientific; Catalog No. A300-500
SYRINGE	**1.0 mL Glass Syringe;** PE Biosystems; Catalog No. 604418. Contains O-rings and ferrules; or other suitable 1.0 mL glass syringe
SYRINGE O-RING	**O-Ring for Glass Syringe;** PE Biosystems; Catalog No. 221102 or other suitable o-ring
SYRINGE FERRULE	**Ferrule for Glass Syringe;** PE Biosystems; Catalog No. 00541 or other suitable syringe ferrule
Taq **DNA POLYMERASE (AmpliTaq® DNA Polymerase)** Store at -20°C.	**5 U/µL;** PE Biosystems; Catalog No. N801-0060
TRAYS	**MicroAmp Base, 10/pkg;** PE Biosystems; Catalog No. N801-0531
	310 Genetic Analyzer 96-well Tray/Retainer Set, 10/pkg; PE Biosystems; Catalog No. 4303864
	310 Genetic Analyzer 48-well Tray; PE Biosystems; Catalog No. 005572
	MicroAmp 96-well Tray/Retainer Set (9700), 10/pkg; PE Biosystems; Catalog No. 403081
	310 Adapter (use with 9700 trays); PE Biosystems; Catalog No. 4305051

TRIS BASE	**(C$_4$H$_{11}$NO$_3$) Trizma base;** Sigma Chemical; Catalog No. T-8524, molecular biology grade

TRIS-HCL	**1.0 M, pH 8.0, 1 L; Trizma base;** Sigma Chemical; Catalog No. S-8524, molecular biology grade

Dissolve 121.1 g Tris base (tris[hydroxymethyl]aminomethane, C$_4$H$_{11}$NO$_3$) in 800 mL filter-purified water. Adjust to pH 8.0 with concentrated HCl. Adjust final volume to 1 liter. Autoclave. Store at room temperature.

TRIS (2.0 M)

Dissolve 242.2 g Trizma base in 800 mL water. Adjust to pH 7.5 with concentrated HCl. Bring to a final volume of 1.0 L with water.

0.2 M TRIS : 2× SSC

Mix 100 mL 2 M Tris-HCl, pH 7.5 and 100 mL 20× SSC. Bring to a final volume of 1.0 L with water.

TAE (20×)	**0.80 M Tris-0.40 M Acetic Acid-20 mM EDTA** **Trizma;** Sigma Chemical; Catalog No. S-8524, molecular biology grade **Glacial acetic acid;** Sigma Chemical; Catalog No. A-0808 **EDTA; (C$_{10}$H$_{14}$N$_2$O$_8$Na$_2$·2H$_2$O);** Sigma Chemical; Catalog No. E-5134, molecular biology grade

To make a 1 L stock solution, dissolve 96.6 g Tris base in 22.8 mL glacial acetic acid and 40.0 mL 0.5 M EDTA, pH 8.0.

10× TBE	**0.89 M Tris-0.89 M Boric Acid-20 mM EDTA** **Trizma;** Sigma Chemical; Catalog No. S-8524, molecular biology grade **Boric acid** (H$_3$BO$_3$); Sigma Chemical; Catalog No. B-6768 **EDTA** (C$_{10}$H$_{14}$N$_2$O$_8$Na$_2$·2H$_2$O); Sigma Chemical; Catalog No. E-5134, molecular biology grade

To make a 1 L stock solution, dissolve 108 g Trizma base, 55 g Boric acid, and 8.3 g EDTA in 700 mL of ddH$_2$O. The pH should be 8.3±0.3. If the pH is out of this range, prepare the solution again. Do not adjust the pH as a change in the ion concentration will affect the migration of the DNA through the gel. Store at room temperature. Discard if a white precipitate is present.

0.5× TBE	0.045M Tris

To make a 1 L stock solution, combine 50 mL 10× TBE with 950 mL ddH$_2$O. Store at room temperature.

0.5× TBE - 0.5 µg/mL EtBr	0.045 M Tris - 0.5 µg/mL Ethidium Bromide, 1 L

Combine 50 mL 10× TBE and 100 µL Ethidium Bromide (5 mg/mL), and bring to 1 L with deionized water. Store at room temperature.

TE (with 0.1 mM EDTA)	10 mM Tris-HCl - 0.1 mM EDTA, 1 L

Add 10 mL 1.0 M Tris-HCl, pH 8.0, and 200 µL 0.5 M EDTA to 990 mL filter-purified water. Autoclave. Store at room temperature.

TE (with 0.5 mM EDTA)	10 mM Tris-HCl - 0.5 mM EDTA

To make a 1 L stock solution, add 10 mL 1.0 M Tris-HCl, pH 8.0 and 200 µL 0.5 M EDTA to 990 mL sterile ddH$_2$O. Autoclave. Aliquot into sterile, plastic tubes when needed, and discard unused portion of aliquot. Store at room temperature.

TEMED	[N,N′,N′-Tetramethylethylenediamine, (C$_6$H$_{16}$N$_2$)] Sigma Chemical; Catalog No. T-9281

Store at room temperature.

TERG-A-ZYME®	Alconox, New York, NY

TISSUE HOMOGENIZER	Dounce tissue grinder; Fisher Scientific; Catalog No. K885300-0015.

TRIS-BORATE SOLUTION Polyacrylamide gel buffer	280 mM Boric Acid Titrated With Tris to pH 9.0, 1 L boric acid (H$_3$BO$_3$); Sigma Chemical; Catalog No. B6768

Dissolve 125.9 g Trizma base in about 700 mL H$_2$O with stirring. Add 17.31 g boric acid. Bring to a final volume of 1 liter with H$_2$O. Check that the pH is 9.0. This is the stock Borate-Tris buffer.

For use in the electrophoresis tank, the stock must be diluted 1:10 with deionized H$_2$O to obtain 28 mM Tris-Borate.

TRIS/EDTA/NaCl (TNE)	10 mM Tris-HCl - 100 mM NaCl - 1 mM EDTA, pH 8.0, 100 mL

Add 1 mL 1 M Tris-HCl to approximately 75 mL of filter-purified H$_2$O. To this solution add 0.584 g NaCl and 200 µL 0.5 M EDTA. Stir until dissolved. Adjust the pH to 8.0 with 1.0 N NaOH and bring to a final volume of 100 mL with filter-purified H$_2$O. Autoclave and store at room temperature.

TRIS-FORMATE BUFFER SOLUTION

240 mM Formic Acid Titrated With Tris To pH 9.0, 1 L
Trizma base; Sigma Chemical; Catalog No. T8524
formic acid (HCOOH); J.T Baker Chemical Company; Catalog No. 0129-1

Dissolve 181.50 g Trizma base in about 700 mL H_2O. While stirring, add 10.0 mL concentrated formic acid. Verify that the pH is 9.0. Bring the final volume to 1 liter with H_2O. Recheck that the pH is 9.0.

Use concentrated formic acid that is 90.8% formic acid. The concentration of concentrated formic acid can vary slightly from lot to lot. The following equation can be used to determine the volume of concentrated formic acid needed to produce 0.12 M formic acid as a function of the concentration of formic acid.

$$M = \frac{(1000 \text{ mL})(1.22 \text{ g/mL})(0.908 \text{ g formic acid/g liquid})}{46.0 \text{ g/mole}}$$

M = 24 moles/liter

Having determined the concentration of concentrated formic acid, the volume necessary to obtain 1 liter of 0.12 M formic acid is calculated as follows:

(1000 mL)(0.12 M) = (X mL)(24 M)
X mL = (1000 mL)(0.12 moles/1000 mL)/24 moles/1000 mL

TUBES

MicroAmp Reaction Tubes, 0.2-mL, 2000/pkg; PE Biosystems; Catalog No. N801-0533

MicroAmp 0.2-mL Sample Tubes, 1000/pkg; PE Biosystems; Catalog No. N801-0580

MicroAmp Reaction Tubes and Caps, 0.2 mL, 1000/pkg; PE Biosystems; Catalog No. N801-0540

MicroAmp Caps, 0.2-mL, 200 strips, 12 caps/strip; PE Biosystems; Catalog No. N801-0534

MicroAmp Caps, 0.2-mL, 300 strips, 8 caps/strip; PE Biosystems; Catalog No. N801-0535

GeneAmp 0.5-mL Sample Tubes, 500/pkg; PE Biosystems; Catalog No. 401957

TUBES	**Microtubes, 1.5-mL;** Sarstedt; Catalog No. 72.692.005
	Microcentrifuge, 2.2-mL; Costar; Catalog No. 3214
UREA	**CH$_4$N$_2$O;** Sigma Chemical; Catalog No. U-6504
UVICIDE®	Vangard International, Neptune, NJ, USA; Catalog No. UV3000; Germicidal Lamp Monitor
VIALS (GLASS VIAL KIT)	Alltech Associates, 15 × 45 nm
SPRING INSERTS	Fisher Scientific; Catalog No. 03-340-53B
VIALS (P/ACE™/eCAP)	Beckman Instruments; Catalog No. 727013
10× WASH BUFFER	**BRL ACES 2.0 Wash Buffer I Concentrate;** Life Technologies; Catalog No. 10354-017

BRL ACES 2.0 Wash Buffer I Concentrate is supplied in a 10× concentration and must be diluted to a 2× concentration for use. Store at 4°C.

2× WASH BUFFER	**BRL ACES 2.0 Wash Buffer I Concentrate**

Combine 100 mL of 10× Wash Buffer I with 400 mL of filter-purified water. Store at 50°C.

WONDER WEDGE	Pharmacia Biotech; Catalog No. 80612788
YO-PRO-1 IODIDE	**1 mM in 1:4 DMSO/H$_2$O;** Molecular Probes, Eugene, OR, USA

Index

304